The Healthy Baby Gut Guide

THE HEALTHY BABY GUT GUIDE

Prevent Allergies,
Build Immunity
and Strengthen
Microbiome Health
From Day One

DR. VINCENT HO

GREYSTONE BOOKS
Vancouver/Berkeley/London

Greystone Books Ltd.
greystonebooks.com

Cataloguing data available from Library and Archives Canada
ISBN 978-1-77164-885-1 (pbk.)
ISBN 978-1-77164-886-8 (epub)

Cover and text design by Fiona Siu
Cover photograph by Poravute/iStock

Printed in Canada on FSC® certified paper at Friesens. The FSC® label
means that materials used for the product have been responsibly sourced.

Greystone Books gratefully acknowledges the Musqueam, Squamish and
Tsleil-Waututh peoples on whose land our Vancouver head office is located.

Greystone Books thanks the Canada Council for the Arts, the British
Columbia Arts Council, the Province of British Columbia through the
Book Publishing Tax Credit and the Government of Canada for
supporting our publishing activities.

The content presented in this book is meant for informational purposes only.
The purchaser of this book understands that the information contained within
this book is not intended to replace medical advice or to be relied upon to
treat, cure or prevent any disease, illness or medical condition. It is understood
that you will seek full medical clearance by a licensed physician before making
any changes mentioned in this book. The author and publisher claim no
responsibility to any person or entity for any liability, loss or damage caused or
alleged to be caused directly or indirectly as a result of the use, application or
interpretation of the material in this book.

CONTENTS

PREFACE

MY STORY

It was a Saturday afternoon and my wife Cindy and I were at a Chinese yum-cha restaurant with our extended family, including our six-month-old daughter Olivia. The dumplings had all been devoured and it was time for dessert. One of the delicacies at the yum-cha restaurant was an egg tart. We fed a bit of one to Olivia. We were introducing her to a range of solid foods at the time and thought nothing of giving her a piece. But just a few minutes later my mother-in-law noticed a new rash over Olivia's lips. The rash then spread all over her face, and she started to become wheezy. She looked quite uncomfortable, and I thought that she might be in pain. We all rushed over to her—myself, Cindy, my brother-in-law, his girlfriend and both grandparents. The waitresses even came over to see what was going on. In the middle of the cacophony, I heard a voice yell out: "Is there an EpiPen?" That panic-stricken voice was mine.

I am a qualified medical specialist, but in this situation I was absolutely powerless. I had no access to any medical resources, and there was no EpiPen in the restaurant. For a few terrifying moments, I watched as my daughter's reaction escalated, my own heart pounding as my worst fears as a parent were realized.

We rushed her into the car to take her to the nearby hospital, but mercifully her reaction began to subside. As we drove, she did not become short of breath nor did her face swell up. When we got to the emergency department, Olivia was assessed as stable and discharged with advice to follow up with an allergy specialist.

Without a doubt, this was one of the most harrowing and distressing events in my life to date. If you are reading this book, you may also have experienced such an event. Your child may have a severe allergy or a mild one. Maybe you have an allergy yourself and are concerned about your kids' risk of developing one. Perhaps you are planning a family, pregnant or hoping to expand your family, and you are aware of the increasing rate of allergies and want to know what you can do to reduce the likelihood that your baby will have one. Whatever the case may be, as a doctor and a parent I can sympathize with the feelings of fear, anxiety and disempowerment that you are probably feeling in the face of the scourge of allergies. I'm here to tell you that you are not alone. I've written this book to provide you with an arsenal of information about the practical things that you can do during pregnancy and birth, and on into your baby's infancy and early childhood, to help prevent allergies developing. A new wave of science has brought fresh understanding about the link between allergies and the gut that every parent concerned about allergies should know about.

ABOUT ME

My dad worked at Westmead Hospital in Sydney, Australia, as a lab scientist, and I remember visiting his lab a few times as a kid, so you might say that it runs in the family. I was entranced by what he did and knew that I was meant to follow in his footsteps.

After high school, I attended medical school at the University of New South Wales. I remained intrigued about lab work and had the opportunity to do a summer research scholarship at the Children's Cancer Research Institute located nearby, working on leukemia cells. This led to a curiosity about cancer cells, which I maintained during my PhD when I studied the biological markers of rectal cancer.

I'm now a senior lecturer at Western Sydney University in Sydney, where I also head a research lab. We're carrying out a different line of research to my previous work on cancer, but there is always a "translational aim," which means that we try to translate a discovery in the lab into an application for the bedside. Practical implementation is at the forefront of my work.

In addition to being a researcher, I'm a practicing gastroenterologist who sees patients with all sorts of gut problems, such as irritable bowel syndrome, celiac disease and inflammatory bowel disease. One thing I've learned over the course of my career is that for many of my patients their problems have been around for a long time, often starting in childhood.

Why did I get into this area of the infant gut? I have a personal interest stemming from my two young children—Olivia, who is four and a half at the time of writing this book, and Brandon, who is two and a half. Looking after their health is my top priority; in light of Olivia's allergy, I couldn't be more motivated to uncover the causes and investigate treatments for allergies. I have devoted countless hours over the last few years to learning the science of the infant gut and allergies—I'm passionate about unlocking the mysteries of childhood allergies and the way that the gut functions.

ABOUT THIS BOOK

My aim is to help you understand the science of allergies and the factors that we believe cause allergies, as well as to offer some sensible and practical strategies for reducing the risk of allergies for your child. This book will also cover managing your child's condition if they have one, and preventing it from becoming worse. We'll investigate the link between allergies and your baby's gut, how the gut develops, what an allergy is and isn't, as well as what you can do to best aid your baby's health during pregnancy, birth and breastfeeding. The first 1000 days of life starts at conception and is a crucial period for parents to take action in allergy prevention. You'll see that each chapter includes specific, practical advice on what you as a parent can do during this time.

In this book, we'll also touch on how hygiene measures might have an impact on childhood allergies. I want to reassure parents that the advice in this book can be followed safely even during times of viral outbreaks such as coronavirus (COVID-19).

Most importantly, we'll cover why a baby's gut health is vital to the prevention of allergies later on in life, so that you as a parent can understand how best to protect and nurture your child's health.

CHAPTER 1

THE SCIENCE OF ALLERGIES

WHAT IS AN ALLERGY?

An allergy is a classic case of mistaken identity. The body confuses a harmless foreign substance with a harmful one, and launches an attack as if the body were under threat. The harmless thing that your child is allergic to is called an allergen. The first time that your child is exposed to a particular allergen, their body produces an antibody—a protein that destroys harmful substances. The antibody that is produced to fight an allergen is called IgE, and after the first contact with the allergen, IgE (allergy) antibodies essentially put your child's body "on call," ready to fight a subsequent exposure to the allergen. Your child's immune system is now sensitive to the allergen; when they come into contact with it again, the IgE (allergy) antibodies are able to organize an attack very quickly, releasing a chemical called histamine which causes the allergic reaction. It is the histamine that can result in itchiness, wheezing, an accelerated heart rate and a drop in blood pressure—the kind of symptoms you would take "antihistamines" to treat.

ANAPHYLAXIS

When a rapid release of chemicals in the body results in a profound and life-threatening response, we call this anaphylaxis. Symptoms can start within seconds or minutes, or there can be a delayed response. Lots of the early signs and symptoms are non-specific and you could see them with any allergic reaction, but if you notice the late signs and symptoms, you should seek urgent medical help or an adrenaline auto-injector such as an EpiPen.

The EpiPen (for adults and older children) or EpiPen Junior (usually for children less than five years old) is a pen-shaped device containing adrenaline. Adrenaline, also known as epinephrine (hence the name EpiPen), is a natural hormone produced by the body in response to stress—you've probably experienced an "adrenaline rush" in your own life during an exciting, threatening or risky situation.

The adult EpiPen has a yellow label and the EpiPen Junior has a green label. They work in exactly the same way—delivering a single, premeasured dose of adrenaline into the thigh muscle. Once the adrenaline is in the body it works to improve blood pressure and redirect blood to vital organs; relax the airways, making it easier to breathe, and stimulate heart receptors to enable the heart to beat more quickly. Adrenaline is essential for fighting back against the following late stage symptoms of anaphylaxis.

Anaphylaxis stages	Signs and symptoms
Early	• nausea • vomiting • dizziness • cough • diarrhea • itchiness • chest tightness • skin reactions, such as raised red bumps (hives) and itching
Late	• weak and rapid pulse • low blood pressure (hypotension) • constriction of the airways • swelling of the tongue, face or throat that can lead to difficulty breathing • wheezing • confusion • unconsciousness

THE RISE OF ALLERGIES

Allergies are a huge public health issue in North America. Hay fever (allergic rhinoconjunctivitis) is thought to affect 10 to 30 percent of Americans and 20 to 25 percent of Canadians. According to the Canadian Dermatology Association, an estimated 17 percent of Canadians will suffer from eczema at some point in their lives, while in the United States that figure is just over 10 percent. Asthma is the third most common chronic disease in Canada, affecting over one in ten Canadians in 2018, while in the United States it affected one in every thirteen persons in 2019. Food allergy is also a big problem, with one in thirteen Canadians having at least one food allergy in 2010–11.

We also know that it's kids who are experiencing the most severe allergies. The leading cause of anaphylaxis in children is food allergies. In North America, from 2010–15, one in 100 children who were admitted to pediatric intensive care units with anaphylactic shock died. Food allergies, therefore, are a serious problem.

It's difficult to work out just how common food allergies are in children, because their reactions to food are often reported by parents without any medical tests being done. If we go by the data supplied by parents on their children's food reactions, in the United States around 6 percent of children in 2016 had at least one food allergy. That figure would certainly be higher now and is thought to be around one in every thirteen children.

Australia, where I live, has the highest rate of confirmed food allergy in the world. The HealthNuts study, based in Melbourne, assessed over 5000 infants and in 2011 found that 9 percent of one-year-old Australian babies had an egg allergy. The good news is that 80 percent of babies will outgrow their egg allergy within a few years. Even those who suffer from severe reactions are still likely to outgrow an egg allergy, with only a very small proportion keeping the allergy for life.

A peanut allergy, on the other hand, is almost the opposite— only 20 percent of children grow out of their allergy. Peanut allergies are the most common food allergy for children in North America, affecting two out of every 100 children. In North America, peanuts and tree nuts account for most of the fatal causes of anaphylaxis in children.

The allergy problem is only going to get worse. The prevalence of peanut allergies in the United States more than tripled between 1997 and 2008, with a 2017 study finding another 21 percent increase since 2010. The World Health Organization predicts that by 2050 an astounding half of the world's population will have had an allergic disease.

From my own experience on the ground, I know that the situation is bad. Waiting times for allergy clinics at public hospitals in Australia have blown out to many months, and even in the private sector there are huge waiting lists for testing, diagnosis and treatment. Many of my immunology colleagues have told me that they already have enough work to keep them going until retirement, as there is no shortage of children needing testing and ongoing management. Many of these children will become adults with chronic allergies.

My North American colleagues assure me that they deal with the very same issues. The waiting lists for allergy clinics in the United States and Canada can stretch out to many months, if not years.

WHAT CAUSES ALLERGIES?

Are we just not dirty enough anymore?

I feel pretty lucky to have had no allergies or major medical issues in my life so far. My mom often reminds me that when I was a toddler I was always playing around in the dirt in our backyard. Apparently I even used to eat ants. Mom believes that contact with "dirt and bugs made you strong," and I've always wondered if there might be some truth to that.

Scientists have proposed the hygiene hypothesis as a way to explain the rising prevalence of allergic diseases. This is the theory that early-childhood exposure to particular microbes (germs) protects kids against the development of allergic diseases. Lifestyle changes in industrialized countries—more sterile urban environments; more time spent indoors—have resulted in less exposure to microbes, especially for children. This has meant fewer infections, but it is also associated with a rise in allergies. In other words, we're not being exposed to the kind of bacteria that we once were, so our bodies aren't learning how to fight certain diseases. It seems like my mom was right—playing in the dirt as a kid is good for you.

The story of the hygiene hypothesis is a fascinating one. In 1989, Professor David Strachan, a London epidemiologist (a scientist who studies diseases within specific populations of people), published the results of a survey of more than 17,000 British children exploring the increased incidence of hay fever in post-war Britain. A curious pattern emerged in the data: the more elder siblings a child had, the less likely he or she was to develop eczema by the age of one and hay fever by the age of 23.

Professor Strachan believed the older children were passing some protective effect on to their younger siblings, and that this special protective effect was exposure to microbes. This early-childhood exposure to particular microbes affected the development of the younger siblings' immune systems, which protected them against allergies.

The catchy title of Professor Strachan's paper was "Hay fever, hygiene, and household size." Funnily enough, the word "hygiene" was only used in the title of the paper and nowhere else in the text. Despite this, the media and scientific community zeroed in on the idea that we're simply not dirty enough anymore. But how exactly does this work? And why are some microbes "good" for us but not others?

How does the hygiene hypothesis work?

The body contains special cells that are activated when we are exposed to a microbe. These highly specialized defender cells are tailored to different microbes. If your body is infected with a particular microbe, only the cells that recognize it will respond. These specific cells then rapidly multiply to fight the infection. Some cells can remember the microbe that previously caused an infection and are able to fight it off very aggressively and quickly, which means that you're unlikely to get sick from that microbe again—you're immune to it now.

Th1 and Th2 immune responses

In the face of an attack by a microbe, the immune system kicks into gear. To combat the infection, immune cells called T helper cells step up to the mark, producing hormonal messages to coordinate the body's immune responses.

There are two types of T helper cells—Th1 and Th2—and they have different functions. Th1 cells are pro-inflammatory and are specially designed to kill off viruses and foreign bacteria that can get inside our cells. But an excessive pro-inflammatory response is going to cause a lot of tissue damage, so there has to be a way to counteract it.

Th2 cells balance the Th1 response by being anti-inflammatory—makes sense, right? But unfortunately, a strong Th2 response can also activate the production of IgE antibodies (remember those?) and cause an allergic reaction.

There is an incredibly delicate balance between Th1 and Th2 activity in our bodies. So if we have reduced contact with microbes, then we'll have less Th1 activity in the body—because Th1 fights foreign bugs. This reduced Th1 activity means that the balance is off kilter, and there will be too much Th2 activity—and therefore, too much allergic reaction-causing IgE in the body. And with this comes a higher risk of allergies.

Many scientists believe allergies are a result of our immune system having overactive Th2 activity. Researchers are looking at ways to bring back the balance—in other words, to encourage the immune system back towards a Th1 response in the hope that this helps to reduce the incidence of allergies. But there is still quite a bit of research to do before these sort of interventions are trialed in humans.

Problems with the hygiene hypothesis

While the hygiene hypothesis explains a lot of things quite neatly, there are a few things that just don't add up, and it has been criticized for providing an incomplete account for the rise in allergy worldwide. Th2 overactivity will settle in time for the majority of children, but some kids continue to have elevated Th2 activity into adulthood. Consequently, the reasons for the rise of allergies worldwide are more complex than just lack of exposure to microbes in early life, or problems with Th1/Th2 balance.

The hygiene hypothesis has been further challenged by results from large studies in Denmark, Finland and the United Kingdom that found no association between the number of viral infections during childhood and allergic disease. Exposure to disease-causing microbes does *not* appear to prevent allergies. In fact, exposure to childhood viral infections can increase the risk of developing asthma for children with a family history of the condition.

Most researchers would agree that the term "hygiene" is a misleading one for the public. The hygiene hypothesis doesn't mean that you should reduce personal cleanliness—it is not an excuse to stop showering or washing your hands! But it is important to know where our microbes come from, how they're spread and how they might help or hinder us.

The "old friends" theory

In 2003, Professor Graham Rook and his colleagues from the University College London posed a new theory which has become popular within scientific circles. Essentially, he restated the hygiene hypothesis to suggest that early and regular exposure to harmless microbes—those that have been with us throughout human evolution, which Professor Rook labels affectionately as "old friends"—trains the immune system to respond to threats. In other words, for our kids to have a healthy immune system, they need to be exposed to a diverse range of bacteria, fungi and other microbes (that don't make us sick) in the natural environment. Through our exposure to "old friends," the immune system learns to function normally and to correctly identify which microbes are friends and which are foes.

In developed countries, exposure to the natural microbial world is much more limited than it once was. We have cleaner sanitation, food and water than ever before. But the rise of allergies has been generated by more than just our sterile living conditions. Widespread use of antibiotics, an increase in cesarean sections (C-sections) worldwide, a reduction in breastfeeding and less time spent outdoors in green spaces also mean less exposure to our "old friend" microbes.

Are allergies genetic?

There is unlikely to be a single gene responsible for all allergies. That being said, genetic inheritance is extremely important when it comes to allergies. A recently published study in the *Journal of Allergy and Clinical Immunology: In Practice* looked at risk factors for food allergy in 80 pairs of twins. The study found pre-existing eczema to be a significant risk factor, but it also concluded that genetic factors played a major role in the development of food allergies. This genetic tendency towards developing an allergy is called atopy.

Researchers have discovered that if you have an older sibling who already has an allergy, then you have a one in three chance of developing an allergy too. If one of your parents is allergic, then you have a one in two chance. If both of your parents have allergies, then your future risk of developing an allergy rises to an astounding 80 percent chance.

Th2 activity, newborns and atopy

The Th1 inflammatory response carries with it a risk of miscarriage. In order to reduce this risk during pregnancy, a pregnant woman's body has an elevated Th2 state to counteract the Th1. So babies are born with the balance between Th1 and Th2 "skewed" towards the Th2 immune response, an imbalance inherited from their mothers. But it is thought that exposure to good microbes in early life can *switch off* this Th2-heavy immune response.

After birth, Th2 activity quickly dampens down in most babies, and they typically go on to show no evidence of allergy. Elevated Th2 activity can persist for a few years in some babies, but it eventually calms down. However, research has shown that babies with a high genetic risk of allergy also have a higher Th2 level. This tends to persist throughout their childhood and can result in these children producing more IgE antibodies when they do have an allergic reaction—which means they're more likely to get severe allergic reactions, such as anaphylaxis. We also know that when children with a family history are exposed to a low level of the allergen, their bodies don't develop immune tolerance (non-responsiveness to the allergen) the way other kids do, putting them at a much greater risk of developing allergies.

However, DNA is not destiny. Just because there is an increased genetic risk does not mean that a child is doomed to a life of allergies. If you or the other biological parent of your child have an identified allergy, then statistically your child has a greater risk of also having an allergy. But if you know that your child has an increased risk, there are things you can do to manage and mitigate those risks.

Known high-risk factors for allergies in your child

A biological parent or sibling with an existing case or history of hay fever, asthma, eczema or food allergies

Severe eczema as a baby

Environmental factors

Our environment may have more to do with the rise in allergies than we realized. We have more exposure to noxious substances—such as pesticides, solvents and air pollutants—in the modern urban world than ever before. Many of these pollutants are known to decrease the Th1 response and enhance the Th2 (allergic reaction) immune response.

The body needs the Th1 and Th2 cells to be in balance. Cigarette smoke decreases Th1 activity and is linked to higher rates of allergy in children, including—unsurprisingly—asthma. Humans who have been exposed to diesel exhaust particles produced more Th2 hormonal messaging cells, supporting a switch towards a Th2 immune response and increased allergies.

Many plastics contain phthalates and bisphenol-A (BPA)—compounds that make plastic more flexible—which can be released into the environment. Phthalates have been shown to induce Th2 cell activity as well as increase IgE (allergy) antibodies, and increased exposure to BPA is linked to a higher risk of developing wheeze and asthma. You might have noticed that the use of BPA-free water bottles has become more popular in recent times; although we're awaiting more data on BPA, I'd say that it's probably better for your baby to drink from BPA-free bottles.

Organic solvents are found in many of the things we use every day, including paint, cleaning products, glue, cosmetics, furniture polishes, tires and construction materials. A group of researchers from Leipzig University and the Helmholtz Centre for Environmental Research in Leipzig looked at whether these organic solvents were risk factors for allergy in early childhood. They found that children exposed to higher amounts of indoor solvents had an enhanced Th2 immune response, and were much more likely to react to milk and egg whites.

Herbicides and pesticides have strong links to allergies. A large population-based study of over 4000 school-aged children in southern California revealed that exposure to either pesticides or herbicides, beginning in the first year of life, was associated with a noticeably elevated risk of early-onset and persistent asthma. In the case of exposure to herbicides, the risk increased 4.5-fold.

WHAT YOU CAN DO AS A PARENT

- Be aware of the hygiene hypothesis—the idea that early-childhood exposure to particular microbes protects against developing allergies—but know that it's not the full picture.
- The hygiene hypothesis does NOT mean that we should neglect personal hygiene. Keep washing those hands!
- Be aware of environmental agents that can increase allergy risk, such as herbicides/pesticides, BPA and organic solvents.
- Understand the hereditary factors that influence the likelihood of your baby developing an allergy.

CHAPTER 2
ALLERGIES AND THE GUT

For our kids to have a healthy immune system, they need to be exposed to a diverse range of bacteria, fungi and other microbes in the natural environment. A key part of the immune system is the gut, which houses a microscopic village of microbes called gut microbiota. Having a diverse gut microbiota is important for the development of a healthy immune system in our kids. An unbalanced gut microbiota can result in a range of health problems, including allergies.

A term very similar to "microbiota" is "microbiome," which refers to our microbes plus all their genetic material. It's become a popular term to use because modern tests that pick up the presence of microbes in our gut detect their genetic signature. "Microbiome" and "microbiota" are often used interchangeably. In this book we'll mainly refer to "microbiota."

OUR IMMUNE SYSTEM AND GUT HEALTH

The immune system is responsible for keeping us well and fighting illness when we get sick. Immunology is becoming an increasingly complex medical field as we learn more and more about the mechanisms that underpin our immune system.

The gut plays a key role in our body's immune system. It's actually our largest immune organ: the gut holds 70–80 percent of the body's cells that are dedicated to fighting illness. Its importance to the immune system is a key reason why the health of your child's gut influences your child's risk of developing allergic diseases later on in life.

THE INFANT GUT AND ALLERGIES

When a baby is born, their gut has no microbes (germs). Microbes enter a baby's body and set up camp inside their gut immediately after birth in a process called colonization. Researchers now believe that allergies can develop if the natural process of establishing a colony of microbes in the baby's gut is altered in some way. We're going to explore this theory step by step.

The weird and wonderful world of gut microbes

Some people think of microbes and shudder. After all, many of us are brought up to believe that microbes are bad for you. We're taught at an early age to wash our hands after going to the bathroom and to cover our mouth when we cough so that we don't catch germs or spread them to other people—sensible and important advice. But microbes are more than just germs that make us sick. It seems that gut microbes may actually be helping to keep us well.

Recently, there has been an explosion in interest in gut microbiota—the village of germs in our gut that includes bacteria, fungi and viruses. Bacteria are by far the most abundant microbes in our gut. There are approximately 40 trillion bacteria in the average adult gut—which spookily is around the same number of cells that we have in our body.

During the first 1000 days of life, a baby's gut microbiota is largely being set. This is a crucial time when you can take action to ensure that your child's microbiota develops properly. Scientists have found that, during the first 1000 days of life, there are three distinct phases of the process by which microbes invade the gut: the development phase, the transitional phase and the stable phase.

At the end of the first 1000 days of life—when a child turns two—the gut microbiota has become stable and more closely resembles that of an adult. Nutritional choices in the first 1000 days of life are therefore hugely important, potentially shaping a child's immune system and deciding whether or not they end up with a particular allergy later on.

Researchers now believe that increases in allergies are due to alterations in the "usual" process of colonizing babies' guts (for example, from the effects of antibiotics). We know from research that there are notable differences in the gut microbes of babies with allergies compared to those of healthy babies. There is less gut microbiota diversity in babies suffering from eczema compared to that of healthy babies. So if the rise of allergies is due even in part to changes in gut development and microbiota in early life, then it's important that we try to understand the infant gut better. The development phase of microbial colonization, which occurs when a baby is between three and fourteen months old, is especially interesting—it's during this phase that a parent can do the most to reduce the likelihood of their baby developing an allergy. We'll be talking about this in more detail in Chapter 9, where I show you a plan to reduce food-allergy risk in your baby.

You might now be wondering when microbes first invade a baby's gut. Is it at conception? In the womb? At birth? When they start solids? Is it already too late for my child? Let's go step by step through your baby's development and see when you can start taking action.

The sterile womb theory

In order to determine *when* the microbes first invade the gut, we need to start at the beginning, with the womb. Are there microbes in the womb? It's an interesting question. For hundreds of years, the womb was thought to be *sterile*—meaning doctors and scientists believed that there were no microbes in the amniotic fluid, unborn baby's gut and placenta.

In 2008, Esther Jiménez and her colleagues from Complutense University of Madrid published the results of their study on the meconium (the delightful name for a baby's very first poo) of 21 term newborns. Meconium was considered a good surrogate measure for fetal gut contents. All meconium samples were taken within the first two hours of each baby's life. Using culture methods, all of the meconium samples revealed the presence of bacteria. Multiple studies have since had similar results, but doubts have been raised around the findings. Even in a study as well designed as Jiménez's one, where the meconium was collected within the first two hours of each baby's life, it is certainly possible that microbes could contaminate the sample given that the newborns had been in contact with the outside world during those two hours. A number of studies, using cutting-edge scientific methods, have provided more information on the different types of bacteria identified in the meconium. But crucially, they couldn't tell if those bacteria were alive or dead.

Researchers challenging the notion of a sterile womb have proposed that the bacteria can get to the placenta and unborn baby's gut via a few different routes: microbes could travel to the womb through the vagina or be carried by cells from the mother's mouth into the blood and then to the placenta. They believe that the mother's placenta has defenses to kill those bacteria, so any bacteria that makes its way into the womb or unborn baby's gut is dead.

My own conclusion, after weighing up all the facts, is that the womb *is* sterile and there is no normal microbiota in the gut of an unborn baby. However, to me, this is not necessarily a bad thing.

Establishing that your baby has no gut microbes in the womb means that childbirth and early contact with the environment is mostly going to dictate which microbes colonize your baby's gut. This means that you as a parent have power over the microbes that *do* end up in your child's gut. Although a child begins with a blank slate in the womb, the right interventions and simple lifestyle steps can actually shape the health of your baby even *before* they're born.

CHAPTER 3
PREGNANCY

So, if the womb is sterile, with no microbes present, what can an expectant mother do to help reduce their baby's risk of allergies? There is a lot to consider with regard to the immune system, pregnancy and allergy risk.

MISCARRIAGE

Part of my passion for working with families and parents-to-be stems from a deeply personal tragedy. For readers who have also experienced a miscarriage, this story may be distressing. I have permission from my wife, Cindy, to share it with you.

In 2015, Cindy and I started trying for our first child, and when Cindy showed me the positive pink strips that confirmed she was pregnant, I was overjoyed. We were going to be a family, and I couldn't wait. We were so thrilled that we told a few close friends and family members straightaway.

Cindy booked an early ultrasound for week ten of the pregnancy. Unluckily, at the exact time of the scan, I had to give my last lecture of the term to my first-year medical students. It was a live, interactive, multiple-choice revision quiz. Though I desperately wanted to attend the scan, I couldn't see an option to reschedule the live assessment that they had been studying for all term. Cindy's parents also had another commitment at the scan time, so Cindy went to the appointment on her own. She assured me that she would be fine, as it was only a screening check. We knew that only later in her pregnancy would we be able to get any good 3D images of our baby.

During the live quiz, my phone was on silent but could still receive texts. Halfway through the lecture, I was surprised to receive a message from Cindy: *Can you call me please?*

I texted back: *In the middle of the lecture. Will call you at the end.*

Thirty seconds went by, and then the text that I will always remember came through. *There's no heartbeat on the ultrasound.*

This happened in real time in front of 120 medical students. I don't know how I got through the rest of that lecture.

Since that time, Cindy has never had an ultrasound without someone else present—either me or her parents. We were both devastated by the miscarriage, and the pain has never completely disappeared.

When I talked to some colleagues at work about what had happened, I was told that there might have been an immune reaction that had caused the miscarriage. No one could tell me much about it, though. So I set myself a goal of learning all I could about the fetal immune system.

THE FETAL IMMUNE SYSTEM

Not so long ago, scientists thought that both the fetal and infant immune systems were simply immature versions of an adult one. In many general medical textbooks, the concept that the fetal immune system is non-responsive to outside threats, such as invading germs, was taken as a given.

Nowadays, we know that this thinking is completely wrong. Medicine is traditionally quite conservative in its approach, and it can take decades before we see shifts in opinion. The new thinking is that the fetal immune system is not immature per se, but actually functions in a different way to the adult immune system.

The fetal immune system is now thought to be responsive to outside threats and therefore able to stimulate immune cells to fight off these threats. But the fetal immune system doesn't normally recognize microbes as foreign threats, because there are no microbes in the sterile womb. However, the unborn baby can incorrectly recognize allergens—which have reached them from the mother—as foreign threats. Encounters with these allergens in the womb are memorized by the fetal system and in some cases the baby can become "sensitized"—i.e. when the baby is exposed to the allergen outside of the womb, their immune system will remember that it is a threat, and this can lead to an allergic reaction.

NUTRITION DURING PREGNANCY

Moms-to-be are under a lot of pressure: from society, about how they should behave as expectant moms; from well-meaning strangers (under the guise of advice); and from their parents, in-laws and partners. However, as my wife Cindy told me when she was pregnant, the biggest pressure often comes from within. Cindy said that a big part of the reason she felt this huge burden on herself during the pregnancy was the constant feeling of guilt that she might not be doing the right thing by her baby, and the interminable stress of trying to work out exactly what the right thing was.

There are endless guides, articles and books written by very knowledgeable people and gurus in their field, so the last thing I want to do is add anything prescriptive to overburden already stressed-out parents-to-be. Instead, I want to present some facts so you can weigh up the evidence and make your own choices.

After delving into the world of gut microbes, I see the fetal gut and womb as empty of living microbes. But there is emerging evidence that the womb and fetal gut may contain allergens (any substance that might induce an allergy). The big question is: can your baby develop an allergic sensitivity to something while they are still in the womb? And, following that train of thought, if your baby can develop an allergy in the womb, should you avoid all potentially allergenic foods when you're pregnant?

Recommended medical guidelines change often, as they rightfully should when new research reveals fresh insights about the body. But these guideline changes can be confusing for pregnant mothers. For example, in 2000 the American Academy of Pediatrics recommended avoiding peanuts during pregnancy, and peanuts, tree nuts, eggs and fish while breastfeeding infants who had a high risk of developing allergy. Eight years later, after new research came out, they revised their guidelines to say that there was now no restriction on what could be consumed during pregnancy or breastfeeding.

In the face of these changing guidelines, I believe that it is important to empower parents to understand the evidence, so they can make informed choices. For parents with allergies themselves, the fear that an allergy could develop in the womb must be a scary prospect.

CAN YOUR BABY DEVELOP AN ALLERGY IN THE WOMB?

Biologically, a mother has an enormous effect on her child's immune system. During the third trimester, the mother begins to transfer a lot of her antibodies to the baby, building up the baby's immune system so that they have some measure of protection once they're born.

Throughout the pregnancy, IgG antibodies are passed from the mother to the baby. IgG antibodies—which protect against infections—are different from IgE (allergy) antibodies and are thought to be the only antibodies that can pass through the placenta. The IgG antibodies provide protection to the infant early in life while they are still developing their immune system. More of the mother's IgG antibodies pass to her baby from early in the second trimester to full term, with the most antibodies reaching the baby in the last trimester. These transferred antibodies generally last four to six months after birth.

While IgG is supposed to be the only antibody that can pass through the placenta, IgE (allergy) antibodies from the mother have been found in the umbilical cord after birth. One theory is that the IgE antibodies "piggyback" on the IgG antibodies during the third trimester, when there is a lot of IgG moving from the mother to the unborn baby.

Allergens (any substance that might induce an allergy) have been shown to pass through the placenta from mother to baby. Food allergens from the mother's diet have been found in the amniotic fluid that surrounds and is regularly swallowed by the unborn baby. This means that it is entirely possible for a baby to develop an allergic sensitivity to something while in the womb, and that the baby is at risk of developing an allergy in the future, especially if they are genetically susceptible to a particular allergy. After all, if both allergen and IgE (allergy) antibodies can pass from the mother to the unborn baby, then shouldn't re-exposure to the same allergen as an infant trigger an allergic reaction?

This thinking may have been partly behind medical recommendations in the early 2000s, which recommended restricting the consumption of peanuts during pregnancy for high-risk mothers (those who have a history of allergies themselves or another child with allergies). The idea of "in utero sensitization" meant that consumption of even trace amounts of peanuts could lead to infants becoming sensitive to peanuts and going on to develop allergies.

WHAT CAN AN EXPECTANT MOTHER DO TO REDUCE THE RISK OF ALLERGIES?

The evidence supports the idea that a baby can become sensitized and therefore susceptible to allergens in the womb, so what can you do during pregnancy to lower the risk?

Visit a farm

You might be surprised to hear this suggestion, but there is strong evidence that a trip to the countryside while pregnant can reduce the risk of allergies for your baby.

The PASTURE study, published in 2012, followed over 1000 children born to farm and non-farm mothers in five European countries. The study was carried out to better understand the causes of childhood asthma and allergies. It showed that exposure to a farming environment during pregnancy is associated with a reduced risk of asthma in children. Why? The thinking is that a farm environment has lots of different germs, microbes and bugs. Being exposed to this kind of microbial diversity during pregnancy improves the baby's immune system and thereby reduces the risk of allergic diseases.

Growing up on a farm has been shown to be protective against allergies in multiple studies (see Chapter 11). Aside from the diversity of microbes, exposure to endotoxins also helps to prevent allergies. Endotoxins are found in the walls of bacteria and are released when bacteria die. They exist virtually everywhere in the environment, including indoors.

Endotoxins are now being extensively studied, because they can switch the immune response away from a Th2 (allergy) pattern towards a Th1 pattern. Endotoxin levels are particularly high in farm stables, where livestock and poultry are kept. It's entirely possible that living on a farm and being exposed to endotoxins, along with microbes, could be extremely helpful in preventing allergies developing during childhood.

However, there's a catch—the exposure is time sensitive, and being exposed to endotoxins later in life seems to *worsen* asthma and hay fever, if you already have them. Research carried out on a rat asthma model by researchers from The University of Western Australia demonstrates this well. A group of rats were sensitized to egg white, and when an endotoxin was given to the rats before or within the first four days of their sensitization, the rats did not develop asthma. But giving the rats endotoxin six or more days after the egg-white allergy was induced caused the rats to develop asthma. The experiment proved that the timing of the exposure to endotoxin, as well as dosage, is hugely important in building immune tolerance.

Pregnancy is an opportune time for early exposure to endotoxins. One study from Jilin province, China, compared the umbilical-cord blood of pregnant women from both farming and non-farming areas. The study also evaluated the mother's endotoxin exposure. The researchers found that the endotoxin content in women from farming areas was much higher than the level in the non-farming group. The cord-blood results from babies born to farming mothers showed that the immune response had already moved away from a Th2 (allergy) pattern and towards a Th1 pattern.

Now comes the really interesting research. The PARSIFAL study was designed to discover the risk and protective factors for allergy in farming children and in kids from Steiner schools—schools with a holistic educational style—in Europe. The children's allergies were assessed by a standard questionnaire, while blood samples were taken to assess IgE (allergy) antibody levels. Researchers also specifically asked about their mother's exposure to farm animals in the questionnaire. A child who lived on a farm and whose family ran the farm was considered a "farm child." Other children were called "non-farm children."

One of the biggest strengths of the PARSIFAL study is the big sample size: 8263 questionnaires were answered. And crucially, researchers found that the prevalence of allergies, hay fever and asthma were significantly lower in farm children compared with non-farm children.

They then evaluated the strength of each of the protective farm effects, analyzing the information they had on the time a mother spent on a farm during pregnancy. The children with the best immune systems were those whose mothers had engaged in stable work during pregnancy! Even if their child was not a "farm child" (in other words, not living on a farm), mothers who did stable work during pregnancy were associated with a decrease in allergen-specific IgE antibodies (that is, antibodies coded and on stand-by to produce an allergic reaction to a particular substance) seen in their children many years later. Interestingly, each additional species of farm animal that the mother was exposed to during pregnancy increased the protective measures that their child's immune system experienced. It seems from this study that exposure to both microbes and endotoxins are vital in building an unborn baby's immune system while in the womb.

So is it time to roll up your sleeves?

Exposing pregnant women to this degree of microbes and endotoxins through "hands-on" contact with a stable or barn goes against everything that we traditionally advise pregnant mothers. In modern medicine, we tend to put pregnant women in a cocoon and are ultra-cautious when advising them on just about anything. I've been guilty of this myself many times, as I'm super wary about the drugs that my pregnant patients take for their chronic gut conditions. I have also spoken publicly about how pregnant women should be careful about consuming certain foods, such as soft cheeses and pre-sliced deli meats, due to the risk of food-borne illnesses. It is a really massive shift in thinking for me to advocate for the possibility that it may be a good thing during pregnancy to be exposed to an environment diverse in microbes and endotoxins. I also have no illusions about how difficult it would be for most women to actually adopt this kind of advice.

As a practical point, if you are planning for a baby in the future and are thinking of having a babymoon on a farm with "hands-on" exposure to a stable or barn, it may be a good idea to get regular exposure beforehand. In other words, test out the waters to see if you can stomach the experience before pregnancy.

Pet ownership and pregnancy

This might surprise you, but there is evidence that having an indoor pet during pregnancy is good for your baby's immune system and reduces the likelihood of developing allergies. The protective effect of pets is less than farm work, but it is still valuable.

The Wayne County Health, Environment, Allergy and Asthma Longitudinal Study (WHEALS) in south-eastern Michigan was designed to examine the relationship between allergies in early childhood and early-life exposure to particular elements, such as pets. Indoor pets were defined as being inside the house for at least one hour each day. In 2008, the researchers concluded that both dogs and cats resulted in a lower IgE (allergy) antibody level in a newborn's cord blood. Three years later, the research team published further results taken from samples over time, and the total IgE antibody levels were found to be consistently lower across the early life of babies whose mothers had kept indoor pets during their pregnancy. So it might be time to get that puppy you've always wanted . . .

Exposure to non-food allergens

You might not think of non-food allergens, such as pollen, as being potentially harmful during pregnancy, but there is evidence that non-food substances can promote allergies. Exposure to non-food allergens in early pregnancy may not necessarily be beneficial—and can even be harmful. A study carried out in Finnish children found that those who had their early gestational period during the pollen season for broad-leafed trees were more prone to food allergies than other children. Research carried out on 387 infants enrolled in the Asthma Coalition on Community, Environment and Social Stress (ACCESS) project in Boston found that the mother's dust-mite exposure was directly related to increased cord-blood IgE (allergy) antibodies.

So what can mothers do about this? One practical thing that expectant mothers can do, especially if they're prone to getting bad hay fever, is to check the pollen count before venturing out of the house. The pollen count refers to the number of pollen particles in one square meter (about ten square feet) of air. In the United States the National Allergy Bureau (NAB) is the section of the American Academy of Allergy, Asthma & Immunology (AAAAI) established to track airborne allergens. The NAB provides pollen counts (trees, weeds and grass) and mold levels from 84 stations throughout the United States and one station in Canada (London, Ontario), which are staffed by AAAAI member volunteers. You can sign up to the NAB for free on their website (aaaai.org) and select your location to receive the latest pollen count.

In Canada, the Weather Network website (theweathernetwork.com) provides very useful information on pollen counts. The data is supplied by the company Aerobiology Research Laboratories and collected from 32 different reporting stations across Canada. You can sign up for a free account and set your location, then click on the "Forecasts and Reports" tab and select "Pollen" to get the day's forecast.

If you're worried about dust mites, studies show that they live in your bedroom more than any other part of the house. Washing your sheets and blankets in hot water at 55 degrees Celsius (131°F) or more will kill dust mites. If you have bedding that can't be washed in hot water, then you can put it in the dryer for at least fifteen minutes at the highest temperature setting to kill off the mites.

In the United States, the Asthma and Allergy Foundation of America (AAFA) in partnership with Allergy Standards Limited (ASL) has developed a certification program for household products that claim to control or reduce exposure to allergens. For example, pillows, mattresses and linen that are considered "dust-mite proof" have to meet certain standards, such as being made of "breathable" material with pores too small to allow dust mites through. By reaching these standards, a product can earn a "CERTIFIED asthma & allergy friendly" mark. In Canada, Allergy Standards Limited (ASL) has partnered with Asthma Canada (AC) to provide the same certification program. The website asthmaandallergyfriendly.com (asthmaandallergyfriendly. ca in Canada) provides a list of certified products.

Stress

Stress during pregnancy has been shown to adversely affect the unborn baby's immune system by inducing it to move towards the Th2 pattern (the allergic reaction pattern in the immune system). Researchers examined the association between mothers' stress levels during pregnancy and cord-blood IgE (allergy) antibodies in 403 infants enrolled in the ACCESS project in Boston. They found increased levels of IgE in cord blood among babies whose mothers experienced higher levels of stress, even when exposed to relatively low levels of dust mites during pregnancy.

It's not always possible to completely reduce stress during pregnancy, of course, but here are some useful tips from my wife Cindy:

- Focus on your baby.
- Spend time with family and friends.
- Get plenty of sleep.
- Eat and drink well.
- Try meditation and yoga.
- Make time for entertainment, such as reading a book or watching your favorite TV series.

Diet

Diet is a big area to cover. It's also one of the most controversial areas, as there's a lot of conflicting evidence as well as the frustration of shifting medical guidelines thanks to evolving research. But diet is a factor over which a pregnant woman actually has a fair amount of control. That being said, eating well during pregnancy can be very challenging when you are faced with morning sickness, an extra-sensitive nose, low energy and cravings. But if you are able to make choices about your diet, you can boost your baby's immune system and lower their risk of allergies.

Fat

A diet with an overall high energy and fat intake can increase the risk of asthma and allergies in your kids. The Irish Lifeways Cross-Generation Study found that higher maternal dietary-fat intake was associated with a greater risk of children with asthma.

A study of over 2000 mothers and babies from Kochi Prefecture in Japan found that mothers with allergic infants had a higher intake of fats and vegetable oil than those with non-allergic infants. Therefore, the general advice is very sound: try not to indulge in too many fatty foods during pregnancy.

Can particular foods cause allergies in unborn babies?

There is a whole range of conflicting data when it comes to particular foods increasing or reducing the risk of allergy. I have to point out that the studies were done on different populations and ethnic groups. This makes it difficult to find consistent data that allows us to make general conclusions for the broader population.

For example, in Germany the LISA study found that a high maternal intake of margarine, vegetable oils and deep-fried vegetable fat was associated with eczema and other allergies in children at two years of age, which is consistent with other studies. But surprisingly, the researchers also found that celery, citrus fruits and raw peppers were associated with eczema and other allergies, too. And the Finnish birth cohort study found that a high maternal consumption of fruits and berry juices was associated with a greater risk of hay fever.

In the Growing Up in Singapore Towards Healthy Outcomes (GUSTO) study, a diet rich in seafood and noodles was associated with a reduced risk of developing an allergy at eighteen months and three years of age. On the other hand, a combined analysis of eighteen European and US birth cohort studies found no evidence that consuming fish and seafood during pregnancy reduced the risk of wheeze, asthma and hay fever in children.

A study in Spain found that sticking to a Mediterranean diet in pregnancy could reduce the risk of wheeze and atopy (the genetic tendency towards developing an allergy) for children at the age of six and a half years. But another study in Mexico found no benefit in a Mediterranean diet during pregnancy with regard to reducing allergic symptoms in children at six years of age, apart from sneezing.

So which study do we believe? When dealing with many conflicting findings, I find it useful to combine the data in a meta-analysis—this helps to resolve uncertainty when studies disagree. The topic of pregnancy diet and allergy risk in infants is an ideal area for a meta-analysis. A very good meta-analysis was published in early 2018 that assessed over 400 studies involving 1.5 million people. Researchers from Imperial College London found that when pregnant women took a daily fish-oil capsule from week 20 of their pregnancy and during the first three to four months of breastfeeding, the risk of egg allergy in infants at twelve months was reduced by 30 percent. They assessed nineteen trials of fish-oil supplements during pregnancy, involving around 15,000 people. The capsules contained a standard dose of omega-3 fatty acids.

The researchers found no evidence that avoiding potentially allergenic foods—such as nuts, dairy and eggs—during pregnancy made a difference to a child's allergy or eczema risk. They assessed other dietary factors during pregnancy, including fruit, vegetable and vitamin intake, but found no clear evidence that any of these affected allergy or eczema risk.

Vitamin D

Another meta-analysis carried out by researchers from the Shanghai Key Laboratory of Children's Environmental Health in 2016 found that lower maternal consumption of vitamin D during pregnancy was associated with an increased risk of childhood eczema. Low levels of vitamin D are common during pregnancy, and deficiencies need to be corrected with supplements. These supplements should be continued for three to six months after the baby is born, particularly if the mother is breastfeeding.

So . . . what should you eat?

The general consensus in the scientific community now is that there is not enough evidence to advise women to avoid specific foods during pregnancy or breastfeeding to protect their children from allergic diseases, such as eczema and asthma. In other words, there is no reason to place any restrictions on an expectant mother's diet—including allergenic foods, such as peanuts, eggs and cow's milk.

The best way forward is simply to adopt sensible, healthy eating habits. I agree with the 2015–2020 Dietary Guidelines for Americans that a healthy, balanced diet rich in fiber, fruits and vegetables is the most appropriate diet for any expectant mother. The guidelines also recommend two to three servings of fish per week with low levels of mercury. The US Food and Drug Administration (FDA) has an excellent web page (fda.gov/food/consumers/advice-about-eating-fish), which shows which fish to eat based upon their mercury levels. Health Canada has similar guidance on consuming fish during pregnancy, which can found online ("Prenatal Nutrition Guidelines for Health Professionals—Fish and Omega-3 Fatty Acids"). If fish can't be consumed, capsules with omega-3 fatty acids could be used as an alternative.

In a full 180-degree flip away from the concept of restriction, some researchers argue that we should be actively encouraging moms to eat a wide variety of foods, including allergens. Unless there is a good reason, such as having a pre-existing food allergy, mothers can and should eat all common allergy-inducing foods during pregnancy—studies suggest there's a good chance that early, repeated exposure to allergens will encourage tolerance of the allergen, rather than sensitization.

Allergens and the womb

We previously discussed the idea that traces of allergens could pass through to a baby in the uterus, and "in utero sensitization" could occur. In theory, it is certainly possible that low doses of and infrequent exposure to allergens could trigger sensitization and the development of a Th2 allergic response. The counterargument is that regular exposure to a high dose of an allergen is likely to lead to the child instead developing a tolerance for it.

In 2014, two studies from the United States on nutrition during pregnancy were published. The first was a huge project called the Growing Up Today Study 2 (GUTS2). It found that peanut or tree-nut allergies were significantly lower in children of mothers who consumed *more* peanuts or tree nuts during pregnancy.

The second study evaluated 1277 mother–child pairs in Massachusetts. During pregnancy, information was taken on the mother's consumption of common childhood allergen foods. Infants were followed to mid-childhood (approximately eight years of age) and assessed for the presence of allergies. The study found that mothers who consumed higher levels of peanuts, milk and wheat during early pregnancy were associated with reduced odds of mid-childhood allergies, especially asthma. The researchers believe that the window of time in which women ate these "risky" foods was vitally important—the first trimester of pregnancy is a key time for the development of the unborn baby's immune system and the creation of allergen tolerance. IgE (allergy) antibodies are produced by the unborn baby at week 11 of gestation, so the researchers suggest that early encounters with food allergens via the mother's diet during this critical period of immune system development can lead to *tolerance* rather than sensitization. On face value, their data is compelling. Moms who ate peanuts during their first trimester were associated with a 47 percent reduction in the odds of their baby developing a peanut allergy during their childhood.

There is certainly a need for more research to determine the best time and dosage for introducing an allergen during pregnancy to induce immune tolerance in the unborn baby. The evidence favors eating more of the common allergenic foods early in pregnancy, but the specifics need to be ironed out more thoroughly. While these studies are promising, it may be wise to await further research.

Probiotics

Put very simply, probiotics are live bacteria that can be beneficial for our overall health at any time, but particularly so during pregnancy and breastfeeding stages. A study by researchers from Imperial College London assessed 28 trials of probiotics supplements carried out on 6000 pregnant women. The researchers found that taking a daily probiotic supplement from weeks 36–38 of pregnancy and during the first three to six months of breastfeeding reduced the risk of the child developing eczema by 22 percent. See Chapter 10 for more information on probiotics.

ANTIBIOTICS AND PREGNANCY

It has been convincingly shown that if a mother takes antibiotics during pregnancy, then there is an increased risk of her child developing allergies. Data from a national registry in Finland has shown, specifically, an increased risk of cow's milk allergy after maternal antibiotic use. Large studies from Denmark and the West Midlands in the United Kingdom have shown antibiotic usage increased the risk of childhood asthma, too. The reason for these heightened risks is that taking antibiotics during pregnancy leads to changes in the gut microbiota of the infant, including less diverse bacterial species.

In fact, maternal antibiotic use during pregnancy can increase a child's risk of being hospitalized with an infection by almost 20 percent. The Murdoch Children's Research Institute considered the data of more than 770,000 Danish children, from newborns to fourteen-year-olds. They found an increased risk of infection when antibiotics were prescribed late in pregnancy or when mothers took more than one course. The group most at risk of gastrointestinal infections were children born vaginally. This makes sense, as the antibiotics would have changed the mother's own gut microbiota as well as the germ environment in her vagina, and so the baby's first contact with microbes through the birthing process would have been altered from the norm. (See Chapter 4 for more on birth, microbes and allergies.)

Obviously, if a pregnant woman needs to take antibiotics to fight a serious infection and maintain her health, then without question she should do so. But antibiotic use during pregnancy should be administered under medical guidance. We should all think carefully about our use (or overuse) of antibiotics and what it might be doing to our long-term health, and particularly the health of our children.

WHAT YOU CAN DO AS A PARENT

- Think about having a babymoon on a farm and visiting the stables. Spending time on a farm before becoming pregnant is highly recommended.
- Do not restrict your diet during pregnancy or breastfeeding. There is no evidence that avoiding any allergenic foods helps to prevent childhood allergy. On the contrary, there is emerging evidence that eating more allergenic foods earlier in pregnancy (during the first trimester) may be helpful in reducing the risk of allergy in your baby.
- Get your vitamin D levels tested during pregnancy; if they are low, then take a vitamin D supplement as directed by your health-care professional.
- Omega-3 fatty acids consumed during pregnancy can reduce the risk of allergy in your baby. Aim for two to three servings of oily fish per week.
- Take probiotics (see Chapter 10 for more information).

CHAPTER 4
BIRTH

The current, widely accepted belief is that a womb is sterile, so there are no microbes in an unborn baby's body. Microbes only colonize a baby's gut when the mother's waters break and the baby moves through the birth canal during a vaginal delivery. As the baby passes through the cervix and vagina, they come into contact with microbes that immediately start to colonize the newborn. Basic rituals of motherhood—labor, postnatal bonding and breastfeeding—all have an extra hidden purpose: transferring the mother's microbiota to the baby's formerly sterile gut.

VAGINAL DELIVERIES VERSUS C-SECTIONS

Research has uncovered differences in infant gut microbiota between babies born via vaginal deliveries and cesarean sections (C-sections). For vaginal deliveries, the first major microbial contact for that baby will be in the birth canal. For C-sections, this step is bypassed, so the microbes in the baby gut are quite different.

The very first colonizers of the infant gut are bacteria that thrive in both an oxygen-rich and oxygen-free environment. They quickly consume oxygen and prepare the way for bacteria that can only thrive in an oxygen-free environment. After bacterial colonization, the gut microbial composition for each child is unique—much like a fingerprint.

A key study carried out by Professor Maria Dominguez-Bello and colleagues at New York University School of Medicine found that bacterial communities of vaginally delivered newborns were dominated by bacteria found in the mother's vagina. In an analysis of the gut microbes in babies by the University of Gothenburg, Sweden, researchers found that 72 percent of the early bacterial colonizers of vaginally delivered newborns could be traced back to the gut microbiota of their mother.

C-section infants, on the other hand, have a gut microbiota that resembles that of their mother's skin. The same study from the University of Gothenburg found that just 41 percent of the early bacterial colonizers of C-section newborns could be traced back to the gut microbiota of their mother. For C-section infants, there is also a lower total diversity of gut microbiota during the first week of life.

Baby microbial diversity and allergies

Studies show that gut microbes are required to shape the healthy development of a child. The importance of early gut microbiota in maturing and developing our immune systems cannot be understated. In general, C-sections appear to delay and change the way an infant's immune system develops—increasing their risk of allergies.

At six months old, both vaginally delivered infants and C-section infants have been found to be colonized by almost the same species of microbes. Despite this, the future health outcomes are different for vaginally delivered and C-section babies. What we know from experiments with animals is that the establishment of the early gut microbiota is needed to stimulate the gut and the immune system to mature. Without the presence of the normal microbial population in the gut (bacteria that help but do not harm the gut), there is nothing to assist the immune system to differentiate between helpful non-harmful bacteria and dangerous disease-causing bacteria. In other words, a normal gut microbiota trains the immune system to correctly identify what can harm it and needs a response, and what does not. Remember that an allergic reaction occurs when the body misidentifies a threat. This early training in correctly identifying threats is crucial to the development of an allergy-free child.

C-section infants show increased odds of developing asthma and obesity. There have also been multiple studies showing an increased risk of other allergies, such as food allergies and hay fever. However, a 2018 meta-analysis review of the research by Edinburgh researchers found no increased long-term risk of developing hay fever or eczema in children born via C-section. But it did find that children delivered by C-section had an increased risk of asthma up to the age of twelve years.

In 2018, a Swedish national study of more than 1 million children was published using recorded data from health-care registries. The study found that both elective and emergency C-sections increased the risk of food allergy. They also found that preterm births (at less than 32 weeks' gestation) had a reduced risk of food allergy, with the reason being unclear.

Data from the Dutch KOALA birth study showed that a baby born via C-section is more likely to have *Clostridium difficile* (a bacterium that can infect the bowel and cause diarrhea), leading to an increased risk of wheeze, eczema and sensitization to food allergens in childhood. *Clostridium difficile* is also more common in infants and children who have received antibiotics.

Elective C-sections are a controversial issue in modern society. The rates of elective C-sections are on the rise around the world, and this is now linked to an increased risk of allergies. However, C-sections do reduce the risk of urinary incontinence and pelvic-organ prolapse in the mother, and in many situations a C-section is the best option for a safe delivery. Frequently a mother will not be given a choice in her method of birth, due to medical reasons. The birth of a healthy baby should always be the priority—and besides, a vaginal delivery is not a guarantee that a child will be allergy-free.

BREASTMILK AND BREASTFEEDING

Breastmilk is often described as liquid gold. It is precious because it provides a rich and complete nourishment specifically tailored for your baby. It is strongly associated with a lower incidence of infectious diseases and allergy in infancy and childhood, particularly asthma. The composition of breastmilk changes daily and reflects the baby's nutritional needs but, significantly, is also subject to the mother's diet. This is part of the reason why good maternal nutrition is so important.

Breastmilk contains immune-boosting substances that are essential in protecting the baby from inflammation and infection. These immune-boosting substances include prebiotics, probiotics, cells and antibodies. All of these contribute to the infant's gut immune defenses.

Breastmilk also contains microbes. The breast-milk microbiota contains more than 700 bacterial species and is crucial in the formation of the newborn's first gut microbiota. It is thought that 30 percent of the gut microbiota in breastfed infants comes from their mother's milk. There are a few different schools of thought on how bacteria gets into breastmilk. During breastfeeding, it is true that the skin of both the mother and baby are in contact, and there can be a degree of backflow between the baby's mouth and the nipple. However, most researchers now believe that special cells in the mother's gut are able to pick up live bacteria and transport them to the breast via the lymphatic system. The mother's gut effectively becomes a fully qualified nutritionist for her baby, so it's a double pronged approach—the bacteria both in the mother's breastmilk and from the mother's skin are transferred to the baby's gut when she breastfeeds.

Expressed milk

Unfortunately, the bacteria in milk taken straight from the breast aren't the same as bacteria in pumped breastmilk. In 2019, researchers from Manitoba, Canada, studied the genes of bacteria in breastmilk samples from 393 healthy mothers, three to four months after they had given birth. Much to their surprise, they found that milk taken from breast pumps had more harmful bacteria, while milk taken straight from the breast had a greater richness and diversity of bacteria. There were also more oral bacterial microbes found in milk from the breast.

Many mothers don't have a choice when it comes to expressing milk, and using expressed milk in bottles is still a good option for parents where breastfeeding isn't possible. Steam sterilizers are particularly good at getting rid of any harmful bacteria found on baby bottles.

Practices of breastfeeding

A very large Australian national survey of infant-feeding practices was conducted in 2010–11. Results from this survey showed that most babies (96 percent) were initially breastfed; 39 percent were exclusively breastfed for less than four months; and 15 percent were exclusively breastfed at five months. Adding in the results of mixed feeding, 69 percent of babies were receiving some breastmilk at four months of age, and 60 percent at six months.

What this data tells me is that it can be very challenging for mothers to continue to exclusively breastfeed until the six-month mark. However, at the same time, mothers are aware of the benefits of breastfeeding, and most will try to ensure that some breastmilk reaches their infants up to the six-month mark.

While breastfeeding should always be encouraged, adhering to exclusive breastfeeding until your baby is six months old is very difficult. I know this from firsthand experience, having seen

Cindy struggle to produce breastmilk for both of our children. She wanted to exclusively breastfeed, as she was fully aware of the benefits, but the amount of milk she produced was completely out of her control.

Mothers in Cindy's particular category (those aged 35 and over, with tertiary education) were found in the Australian national survey to have higher rates of starting breastfeeding and higher-intensity breastfeeding for longer periods. However, being knowledgeable does not guarantee successful breastfeeding; for many women, physical issues need to be taken into account, such as simply whether adequate amounts of breastmilk can be produced. Some babies just aren't able to latch, suckle or take in enough breastmilk to gain weight and get enough nutrients. There can also be post-birth complications or other conditions that make breastfeeding painful for the mother and/or stressful for both mother and baby. Breastfeeding is not a given, and it does not come naturally to all mothers and babies. Plus, there are also other challenges, such as returning to work and sharing the feeding load with a partner.

Cindy felt quite a lot of guilt with our first baby, Olivia, when she couldn't produce enough milk and needed infant-formula supplementation. Luckily, she had a really supportive midwife who encouraged her to continue to breastfeed wherever possible but acknowledged that she needed to supplement Olivia's nutrition with formula. This made it a lot easier for Cindy the second time around with Brandon—knowing that milk production was out of her control, and feeling comforted that her baby would still get enough nutrition if she supplemented breastmilk with formula.

Mixed feeding

Mixed breastfeeding (supplementing breastmilk with formula) is different from exclusive breastfeeding, but it does seem to have benefits over formula feeding alone. A research group in Hong Kong studied the gut microbiota of a group of breastfed infants, a group of formula-fed infants and a group of mix-fed infants who consumed both breastmilk and formula. Their findings from these babies aged two to four months suggested that partial feeding with breastmilk could still maintain the major gut community composition as seen in the breastfed group.

The Shanghai Children's Health, Education and Lifestyle Evaluation was a large population-based survey carried out in 26 primary schools in Shanghai, China, in 2014. The mode of delivery (vaginal versus C-section) and the child's history of asthma and hay fever were reported by parents. There were 12,639 children included in the analysis. It found that non-emergency C-sections were associated with increased risks of both childhood asthma and hay fever. Notably though, in children fed exclusively by breastfeeding or by mixed feeding in the first four months after birth, these risks were much lower.

The research suggests that any breastmilk is better than none at all. For mothers who struggle to produce enough milk to breastfeed exclusively in the first six months, mixed feeding is the next best thing—and the health outcomes are not dissimilar down the track.

Infant formula

We know that exclusively breastfeeding to six months is the best practice, and we should be encouraging mothers to embrace this. But where this is not achievable for a mother, through no fault of her own, we need to make replacement infant formula as safe as possible. When I give my own children a bottle containing infant formula, I need to know that it is the best it can be. Ultimately, it's about trying to make infant formula as close to breastmilk as possible.

As I've come to know the infant-formula industry better— I've sat on several advisory round tables for infant-formula companies—I've realized that there have been plenty of good, randomized controlled trials carried out on infant formula. This is a good thing, because high-quality evidence does get incorporated into the guidelines of medical societies and ultimately those of regulatory bodies. In the United States, the Food and Drug Administration (FDA) has specific requirements that govern what can go into infant formula. It even governs how manufacturers can label and package infant formula, and how they can promote their product. This same function in Canada is overseen by Health Canada and the Canadian Food Inspection Agency (CFIA).

There have been tremendous advances in infant-formula development, and the technology used is now incredibly sophisticated. Until the end of the nineteenth century, babies were fed animal milk as a supplement or replacement for breastmilk, most commonly cow's milk. In 1760, Jean-Charles Desessartz discovered that there were differences in the composition of cow's milk and human milk. Desessartz determined that human milk was the best choice for baby nutrition.

Given that human breastmilk was recognized as ideal, even then, it was not long before scientists tried to emulate breastmilk. In 1865, an enterprising German chemist called Justus von Liebig developed the first infant formula. This formula was made up of cow's milk, wheat, malt flour and potassium bicarbonate; marketed as "Soluble food for babies," it was a runaway commercial success. This spurred more development in the field. In 1867, Henri Nestlé (the founder of the Nestlé food and beverage company) came up with his infant formula, which combined cow's milk, wheat flour and sugar. In 1884, John B. Meyenberg patented a process for sterilizing concentrated milk in tin cans.

Unfortunately, by the turn of the twentieth century, poor sanitation had led to many infant deaths; an important cause was bacterial infections from unwashed feeding bottles. Strenuous efforts were then made to educate the public about germs, and improvements were made to ensure that infant formula was safer and more nutritious. Researchers recognized that the cow's milk in infant formula made a few babies quite sick, so in the 1920s, they started to develop non–milk–based formulas for babies with a cow's milk allergy.

During the 1940s and 1950s, consumers in the United States increasingly adopted the primary use of infant formula. Aggressive marketing of infant formula corresponded with a steady decline in breastfeeding until the late 1970s. In 1980, the first Codex Alimentarius standard for infant formulas was produced, outlining a set of nutritional requirements for infants. Breastfeeding rates only started to increase again from the end of the twentieth century, thanks to the public's greater awareness of the health benefits for babies.

Formula problems

The gut microbiota of exclusively formula-fed babies is quite different from that of breastfed babies. The microbiota of formula-fed babies shifts towards an adult's at a quicker rate, with higher overall bacterial diversity. Unfortunately, developing this more complex, adult-like gut microbiota at an earlier age isn't beneficial, as it means that these exclusively formula-fed babies are at an increased risk of eczema. They're also at more risk of becoming overweight.

Parents who exclusively feed their babies with infant formula may be worried when they discover this. As I learned firsthand from Cindy's experience, mothers feel a lot of guilt and stress when it comes to the health of their babies. It can be a physically and emotionally exhausting experience to find, read and follow everything written about making their babies healthier.

The good news for parents with babies on exclusive formula feeds is that there are some simple things that can be done to help reduce the risk of their babies developing allergies. For example, the use of probiotics by pregnant women, breastfeeding women and their babies is just one way to reduce the risk of infant allergies. Another way is to make sure that the baby consumes prebiotics. We'll get to probiotics and prebiotics in Chapter 10, and I'll explain how useful they are for your baby.

WHAT YOU CAN DO AS A PARENT

- If it is at all possible, safe and in the best interests of both mother and baby, have a vaginal birth. Of course, circumstances often dictate that a C-section is needed for the safety of mother and baby, and medical professionals will make this choice on behalf of parents.
- Breastfeeding your baby exclusively to six months is recommended, and breastfeeding can continue until twelve months of age. However, if you're unable to breastfeed your baby exclusively, giving your baby even a small amount of breastmilk is a good thing.
- If you are pregnant or a breastfeeding mother with a baby who has a high risk of developing an allergy, consider taking a probiotic. See Chapter 10 for more information.
- Prebiotic supplementation in non-exclusively breastfed infants is also worthwhile (see page 171).

CHAPTER 5

BABIES, DIETS AND PACIFIERS

The amazing thing about the gut microbiota is that from infancy there is a massive growth in the different types of microbes in the gut. A good diet in the early years of life can make a real difference to the health of a child. There's also surprising new evidence around pacifiers, and what you can do to reduce your child's risk of developing allergies.

DAY 270 ONWARDS

In the first 1000 days of life, the clock begins ticking at the time of conception. Nine months breeze by; when the baby is born, if it is at full term, we are at day 270. The first 1000 days finishes on the child's second birthday.

What happens to the gut microbiota as the child develops?

A child's gut passes through three phases:

- a development phase from three to fourteen months
- a transitional phase from fifteen to 30 months
- a stable phase from 31 months onwards

The first 1000 days of life includes the development and transitional phases. The number of infant gut bacteria increases drastically with the introduction of solid foods. The transitional shift in the infant gut microbiota happens between one and two years of age; after the transitional phase the gut microbiota becomes stable and more similar to an adult's.

INTRODUCTION TO SOLID FOODS

The process of weaning—when solid foods are introduced to your baby—typically starts at around four to six months of age and continues until the child is approximately two years old. The World Health Organization (WHO) recommends a gradual introduction of solid foods while the baby is still taking breastmilk and/or infant formula. The gold standard for breastfeeding is to feed exclusively until six months of age, so this is the ideal time to start introducing solid foods.

Generally, at around six months of age, babies begin to show interest in eating solid foods. If babies are started on solid foods too early, their digestive system will not be mature enough to handle the food. Before four months of age, their pancreatic function, small-bowel absorption and fermentation ability is underdeveloped. It is only at six months of age that the pancreas will secrete enough enzymes to digest the starches and proteins of solid foods.

At around six months of age, infants are also developmentally ready to start solid foods; they can hold their head and neck up straight, which allows food to be swallowed safely. At that age, the "tongue-thrust" reflex—where infants push out anything in their mouths using their tongue—disappears, and they should be able to reach out for food and open their mouth when offered food on a spoon.

Babies are born with significant iron stores in their livers. At six months old, the baby's iron supply from their mother is depleted. Since human milk is quite low in iron, the baby then needs to consume either an iron-fortified infant formula or a diet of iron-rich foods, such as meats, dried peas, beans and legumes, tofu, eggs and leafy vegetables. During weaning, a baby should also be consuming vitamin D–fortified milk, yoghurt and cereals, as vitamin D is necessary not only for building bone strength but also for the proper maturing of the gut microbiota.

As solid foods are introduced into the baby's diet, the gut microbiota begins the process of evolving from a simple environment that digests only breastmilk or formula to an environment that can digest richer, more varied foods. The introduction of solid foods triggers the change in the gut microbe community, but the gradual reduction in breastfeeding has the greatest impact on the baby's gut microbes.

The timing of solid foods and allergies

Research has now determined the best time to introduce solid foods to reduce allergies. A research team from the United Kingdom followed a group of over 1000 infants from birth to two years old and documented suspected allergic reactions to food. They found that the infants who developed food allergies were more likely to have been introduced to solids *before* seventeen weeks (four months) of age. Their findings named seventeen weeks as the crucial turning point. Introducing solid foods *before* seventeen weeks appeared to promote allergies, whereas introducing solid foods *after* this time seemed to promote immune tolerance.

The researchers also found that continuing to breastfeed while introducing cow's milk had a protective effect against allergies. They suspect that the breastmilk's positive role in the baby's immune system is even greater when the immune system is exposed to an allergen such as cow's milk protein at the same time. It's the perfect balance, in other words.

Multiple studies have confirmed that there is no evidence to justify a delayed introduction of solid foods after six months of age to prevent allergies in both high- and low-risk infants. The American Academy of Pediatrics (AAP) guidelines recommend starting solid foods at around six months of age, but not before four months. Breastfeeding your baby while introducing solid foods is also recommended. Guidelines from a number of other national allergy societies and the World Health Organization (WHO) mirror the AAP's recommendation.

Benefits of a diverse diet

At the end of the first 1000 days of life, when a child is two years old, they have achieved a stable gut microbiota. At that point, the effects of prior breastfeeding are no longer seen in the gut microbiota. Diet is the key to shaping the numbers and types of helpful bacteria in the gut, as well as maintaining the right balance of the different sorts of microbes.

Nutrition is critical to regulating the gut microbiota throughout life. There has been extensive research carried out on the role of diet in early life and the flow-on risk of allergy. A striking finding from the research is that eating from limited food groups during the first year of life can increase the risk of asthma and other allergies during childhood. In contrast, eating from many different food groups in the first year of life appears to reduce the risk of asthma, food allergy and food sensitization.

Some studies in the decade leading up to 2012 suggested that the introduction of foods such as fish before one year of age, or an early exposure to cow's milk, could have a protective effect against allergic diseases. But the evidence for the allergy-preventing benefits of a diverse diet for infants didn't become clear until the PASTURE study, a European study of 1133 children that evaluated risk factors and preventative factors for allergies. As part of the study, parents used monthly diaries to report their feeding practices for their babies, who were aged between three and twelve months. They also carried out weekly checks, noting any itchy rashes on their infants. For each food item, parents indicated whether it was given to the child in the previous four weeks and, if so, how often. A diversity score was calculated, which involved defining the major food items—which ones were introduced in the first year of life to 80 percent or more of the children. The six major food items were vegetables or fruits, cereals, bread, meat, cake and yoghurt. The consumption of cake may be culturally specific for this European study, and other populations may not have cake as part of their infant solid-food diet (I'm sure that health experts would have concerns about the sugar content of cake!).

Feeding practices during that first year of life seem to be associated with the development of atopic (genetic tendency to develop) eczema. The investigators showed that the diversity of food items introduced in the first year of life reduced the risk of atopic eczema later in childhood. Introduction of yoghurt in the first year of life had quite a strong protective effect against the eczema occurring after the first year of life.

The investigators hypothesized that exposure to a variety of allergens, such as the food protein from egg, during a specific period early in life could be the key to developing immune tolerance.

The second PASTURE study

The investigators were able to explore this specific time period with another study involving the PASTURE birth cohort, which was published in 2014, two years after the first study. In this second study, foods that were introduced in the first year of life were examined, looking for any association with asthma, food allergy, hay fever and genetic susceptibility towards allergies in 856 children up to six years of age. The feeding practices in the first year of life were included in the study.

They carried out diversity scores similar to the first study, based upon monthly food diaries but in more detail. They looked at the diversity scores for the six major food items introduced in the first six months of life (vegetables or fruits, cereals, bread, meat, cake and yoghurt), then the same six major food items introduced in the first twelve months of life, and finally all food items introduced in the first year of life, which they divided into fifteen distinct groups. This was meant to capture all potentially allergenic foods (cow's milk, yoghurt, other milk products, eggs, nuts, vegetables or fruits, cereals, bread, meat, fish, soy, margarine, butter, cake and chocolate).

The data showed convincingly that a high diversity of foods (including allergenic foods) in children's diets within the first year of life protects them against the development of asthma and food allergy up to six years of age.

When researchers compared the protective effect of food diversity within the first six months against the first year of life, they found that the effect of food diversity was significantly greater in the first year of life. This is highly suggestive that eating a variety of food types—including allergens—between the ages of six and twelve months is the key to protecting against the development of later childhood allergies.

In light of this, I've created a "nine food allergens in nine weeks" plan for you to follow when introducing solid foods to your baby (see Chapter 9). The meal plan is designed to provide a diverse range of foods for the infant in addition to introducing the food allergens. The five main diverse groups of foods in the meal plan have been slightly modified from the PASTURE study for a North American context, and they are:

1. vegetables and legumes
2. fruits
3. grains and cereals
4. high-protein foods (meat, fish, tofu, poultry, eggs)
5. dairy (milk, cheese and yoghurt)

Cake has been removed as a major food group, given its generally high sugar content.

Dietary fiber and allergies

Dietary fiber is a type of carbohydrate that is fermented by the gut bacteria in the colon. There are three main types of fiber that have different functions and health benefits.

Insoluble fiber does not dissolve in water. It softens the contents in the colon and supports more regular bowel movements. Good sources of insoluble fiber are wheat bran, rice bran, the skin of fruits and vegetables, nuts, seeds, cereals, dried beans and wholegrain bread.

Soluble fiber, on the other hand, does dissolve in water. It slows down the emptying process of the stomach to make us feel fuller. Good sources of soluble fiber are vegetables, fruits, seed husks, oats, psyllium, legumes, peas, barley, soy milk and other soy products.

Resistant starch is the third type of fiber, and it is the part of starchy food that resists normal digestion in the small intestine. When starch arrives in the colon, it feeds healthy bacteria that then convert the starch into short-chain fatty acids to help nourish the cells lining the colon. Good sources of resistant starch include undercooked pasta, cooked and cooled potato and rice, raw bananas and legumes.

We know from mouse models that eating a high-fiber diet suppresses the development of food allergies. In one mouse peanut-allergy model, a particular receptor found in cruciferous vegetables such as cabbages, brussels sprouts and broccoli has been found to reduce the likelihood of developing peanut allergy.

In another study, mice fed a high-fiber diet and exposed to house dust mite extract developed less severe allergic airway inflammation compared with mice on a control diet. An increase in short-chain fatty acid levels in the colon and blood was seen in mice on the high-fiber diet, and it is thought that the short-chain fatty acids were acting to suppress the Th2-driven immune (allergic) response.

In young children, dietary fiber has been shown to reduce the risk of asthma and wheezing. A meta-analysis of two randomized controlled trials in children aged two to five years found that prebiotic dietary fiber reduced the risk of asthma or wheezing by 63 percent when compared with the control group.

Rural African children generally consume a low-fat, vegetable-based diet, and they have a more balanced environment of gut bacteria. Additionally, the gut microbiota of rural African children has an abundance of a particular bacterial species not found in European children; this bacteria can break down starchy fiber, a staple part of the rural African diet. These children have a low rate of allergies, which reinforces the importance of dietary fiber as a cornerstone of the infant solid-food diet.

How much dietary fiber should a child consume?

Despite it being a key element of infant diets, human breastmilk does not contain any dietary fiber. According to the Nutrient Reference Values for Australia and New Zealand (nrv.gov.au), there are no daily fiber intake recommendations for the first year of life because of a lack of data on infants and fiber intake.

Even for experienced health professionals, working out how much dietary fiber young children need is difficult to do. Some have tried to calculate how much daily fiber is required for a child based upon their energy requirements; others use a child's body weight. Professional guidelines for older children's dietary fiber intake have been issued by different organizations, such as the Institute of Medicine (based in the United States), American Academy of Pediatrics, European Food Safety Authority and the Australian Nutrient Reference Values. Unfortunately, these guidelines differ wildly, so it's understandable parents are confused.

We now know that children need more daily fiber that what has been previously recommended. Though there is a lack of data in the 6-12 month age group, a Brazilian study found that infants in this age group were able to tolerate up to 10 grams of dietary fiber without ill effects. The study, published in 2011, found that constipated infants benefited substantially from a high-fiber diet containing lots of wheat bran.

The type of dietary fiber is extra important for young children. Data from the US National Health and Nutrition Examination Survey (NHANES) showed that consumption of grains (such as brown rice, oats, wheat, rye) by young children in their first 1000 days of life (up to two years of age) was associated with higher nutrient intake compared to children who did not eat grains. The best type of grains for young children are wholegrains—grains that are not broken down by processing. Young children who eat at least 75 percent of their grains as wholegrains were found in the 2016 US-based Feeding Infants and Toddlers Study (FITS) to be consuming the highest amount of daily dietary fiber. Good examples of wholegrain foods include whole- and mixed-grain breads, rolled oats, wholegrain pasta, brown rice and rice cakes.

Using both the US Institute of Medicine and American Academy of Pediatrics guidelines, I've created a table to show how much dietary fiber a child should be consuming according to their age. It's strongly recommended that young children are given wholegrains as their dietary fiber.

Daily fiber intake

Age	Boys/Men	Girls/Women
6–12 months of age	Up to 10 grams of fiber/day	Up to 10 grams of fiber/day
Age 1–8	AGE + 10 grams of fiber/day	AGE + 10 grams of fiber/day
Age 9 onwards	Double the AGE = grams of fiber per day until the age of 15. Stay on this until age 18.	Double the AGE = grams of fiber per day until the age of 13. Stay on this until age 18.
Adults (18 years +)	At least 30 grams/day	At least 25 grams/day

It's all very well to know how many grams of fiber your child should be eating, but how do you know how much fiber is in the food you're putting in your shopping cart? You can use the

US FoodData Central database, managed by the US Department of Agriculture: fdc.nal.usda.gov. Type in the food item that you want to check in the search bar and it will give you a breakdown of the nutrients, including dietary fiber, in an extensive range of common foods.

TIPS FOR ENCOURAGING YOUR BABY TO EAT A DIVERSE RANGE OF FOODS

For young infants, it is important to make sure that the texture of the food is appropriate for their developmental age. Infants need to develop and refine their eating motor skills, so it's important to change the texture of foods between six and twelve months of age. The table below shows the different textures of food appropriate for babies during the first year of life.

Texture of baby foods in the first year of life

Age	Texture of food	Examples
6–7 months	Blended or pureed	Pureed steamed carrot
7–9 months	Mashed and grated foods, soft lumps and soft finger foods	Mashed pear
9–12 months	Food with lumps that need munching, chopped foods, finger foods, mixed textures	Cooked noodles

Gagging is common for infants, especially when introducing new foods. As a parent, it can be alarming when your child gags, but understand that there is a difference between gagging and choking. Gagging is a protective mechanism, and it doesn't mean that the particular food should be avoided in the future. Babies need to learn how to eat new foods with different textures. But don't push them to eat extra mouthfuls if they've had enough. It's important not to encourage a child to overeat or to finish off everything on their plate. Children need to learn to listen to their own hunger/full signals, and will communicate this to their parents.

Low-fat diets are not suitable for children under two years of age. Your infant will need the fat found in full-fat milk, cheese, yoghurt, custard, avocado, nut products, meat and oily fish for their healthy growth and brain development.

And here's a final note on "fussy eaters." It is quite normal for young children aged between one and three years old to become "fussy" in their eating habits, such as refusing familiar food or being unwilling to try new foods. Toddlers can also experience a reduced appetite, as they are growing more slowly than during their first year of life.

Young children should be encouraged to touch, smell and taste new foods. It may take quite a number of exposures (some studies suggest up to fifteen exposures) for a child to accept a new food. With young kids, it's better to think about what they eat on a weekly basis rather than scrutinizing what they eat every single day. If they have lots of energy, are still growing and not getting sick, then they're okay. Sometimes, when parents worry that their kids aren't eating enough good food, they can feed their own anxiety by trying to "force feed" their kids. Be patient, and try not to stress—kids often figure these things out in their own time.

TIPS FOR INFANTS ON A VEGETARIAN DIET

If your baby is on a vegetarian diet, you need to take care to ensure that they are getting enough fat, iron and other nutrients, such as zinc and vitamin B_{12}.

A general rule of thumb when introducing solid foods is to substitute all animal-flesh foods with vegetarian-protein foods, such as legumes, baked beans, lentils, tofu or nut butters. The use of iron-fortified foods such as iron-fortified cereals is a good option, and eggs can be acceptable for some vegetarian families. Dairy foods such as cheese and yoghurt can be introduced from six months onwards, and may also be acceptable for some vegetarian families.

For young children on a pure vegan diet, a vitamin B_{12} supplement is recommended. To find out more information about a healthy vegan diet for your baby, see an accredited practicing dietitian.

WHAT YOU CAN DO AS A PARENT

- Where possible, exclusively breastfeed your baby until they are six months of age.
- Introduce solids no earlier than four months. Ideally, solids should be introduced at around six months.
- You can continue to breastfeed your baby until they are twelve months of age while introducing solids.
- Ensure that your child has a diverse diet, with the major food groups represented (vegetables and legumes, fruits, grains and cereals, and high-protein foods such as meat or tofu and

dairy). Yoghurt is a great form of dairy for babies and has been shown to reduce the risk of eczema.

- Dietary fiber should be part of your baby's diet and has been shown to reduce the risk of childhood asthma and wheezing. It's strongly recommended parents give young children wholegrains as the main form of their dietary fiber.

- The time between six and twelve months is a critical period when the baby's gut microbiota is developing. Studies have shown that a diverse diet during this time has a substantial protective effect for allergies. Introducing common allergenic foods during this time can help your child develop a tolerance for them.

PREMASTICATION AND PACIFIERS

There is another way in which parents may be able to pass on their microbiota to their babies. Imagine a mother bird feeding her chick a wriggling worm. During that process, food as well as microbes from the mother bird's beak can be passed to the chick.

Humans can do this with their infants, too, using a process called premastication. Mastication means chewing, so premastication is the practice of feeding an infant with food that has been chewed by their mother or other caregivers. This is a traditional practice that begins for the infant as early as one month old, and it is practiced by many diverse traditional societies today. Premastication is very rarely observed in modern Western societies due to the abundance of processed and sterilized infant food. Infants in Western societies also start solid foods when they are between four and six months of age, which is quite a bit later than when premastication usually starts.

Feeding a baby pre-chewed food means that the child is consuming the parent's saliva, which contains microbes. Adult saliva may play an important role in allergy prevention. A number of studies have noted that vaginally delivered infants have a greater diversity of salivary microbiota than C-section children. C-section infants have a higher risk of developing allergies, so research has studied the idea that early exposure to oral microbes for infants could protect them from allergies.

In 2013, the ALLERGYFLORA study was carried out at Mölndal Hospital in Gothenburg, Sweden. Pregnant women were recruited into the study, and their babies were included in the birth cohort one to three days after delivery. Eighty percent of the babies had one allergic parent, and hence these infants were at high risk of developing allergy.

An interview was carried out with mothers that focused on the pregnancy, delivery, family structure and housing conditions. Parents kept diaries that covered the lives of their infants during their first six months of life. During a telephone interview when the babies were six months old, parents were asked about whether their child used a pacifier. If the answer was yes, they were asked if the pacifier was cleaned by boiling or rinsing it in tap water, or whether parents sucked on it to clean it.

A total of 187 infants were included in the study. There was an excellent follow-up rate, with 184 of these babies followed to eighteen months of age, and 174 until they were three years old. Three-quarters of all infants used a pacifier. Interestingly, almost half of the infants who used a pacifier had a parent who cleaned the pacifier by sucking it before giving it back to the infant.

The results of the study were striking. Cleaning the pacifier by sucking it was strongly associated with a reduction in allergy development. For eczema, the relative risk reduction was 73 percent at eighteen months; for asthma, it was 84 percent at eighteen months.

The researchers also wanted to know if there was a difference in the mouth microbes of infants whose parents sucked a pacifier to clean it and those whose parents did not. Bacterial DNA analysis in vaginally born infants revealed that there were distinct groups of oral microbes generated by whether or not the pacifier was sucked clean by their parents. The use of boiled water or rinsed water to clean the pacifier in this study had no effect on allergy risk.

The researchers pointed out that the mouth is the first point of contact between foreign substances and the lymphatic system (the network of tissues and organs that transports infection-fighting white blood cells around the body). There is a lot of dense lymphoid tissue in the tonsils, which contain cells that identify foreign threats. The researchers believe that because there are a lot of important immunity cells in the tonsils, immune tolerance to allergens could be generated in the mouth and passed on to a baby through saliva. Additionally, when we swallow our saliva, mouth bacteria move to the stomach and affect the composition of the gut microbiota.

WHAT YOU CAN DO AS A PARENT

- Clean your child's pacifier with boiled water and then suck on it for a moment before returning it to their mouth—research suggests your mouth microbes will reduce their risk of eczema and asthma.
- But don't suck on a pacifier and give it to your baby if you are feeling unwell, have a sore throat, flu-like symptoms or a dental infection.

CHAPTER 6
COMMON ALLERGIES

Allergies come in a wide range of forms. In this chapter, we will discuss common allergies, how they occur, what to look for and the usual management strategies.

TESTING FOR ALLERGIES

There are two ways that we test for allergies. The first is the skin-prick test, which can be taken by children of any age, including young babies. An experienced clinician uses a small plastic pricker to make a shallow break in the skin (often on the child's forearm or back), and a tiny amount of allergen is dropped into the pinprick. The pricker usually doesn't cause any bleeding. Fifteen minutes after the injection, the size of the welt or raised bump is measured. The test can't be carried out while a child is on antihistamines. Although the test is considered safe, there is a very small risk of anaphylaxis, so an allergy specialist should be around to monitor and provide treatment if needed.

The other way that we can test for allergies is with a blood test, which measures the antibody IgE levels of different allergens in the blood. This can be done while a child is on antihistamines.

Both of these tests are really useful for diagnosing allergies and they each require an experienced doctor to accurately interpret the results.

ATOPY AND ATOPIC MARCH

Atopy is a genetic tendency to develop allergic diseases. A child with atopy will produce specific IgE (allergy) antibodies after they are exposed to an allergen, which means that they're sensitive to that allergen. Not all children with atopy will go on to develop any of the childhood atopic diseases of eczema, food allergy, asthma and hay fever (allergic rhinoconjunctivitis)—it is simply a genetic predisposition.

Atopic march refers to the common progression of allergic diseases that begins in infancy. Eczema is the first allergy to develop in children, and it usually begins in early infancy. It is followed closely by food allergy, and after that comes asthma and finally hay fever. For up to 40 percent of children with eczema, there is a progression to asthma and/or hay fever, and most develop respiratory symptoms before five years of age.

The atopic march is really important for research into childhood allergies, because by knowing the progression of allergic diseases, we are able to plan and test interventions in high-risk children. So, if a child has severe eczema, they will have a high risk of developing a subsequent allergy, such as a peanut allergy. In cases like these, we can test, manage and mitigate to reduce the risk of further allergies. Doctors rarely have difficulty diagnosing atopic diseases.

COMMON ALLERGIES

The big four allergies are eczema, asthma, hay fever and food allergy. Having a biological parent or sibling who currently has or previously had any of these four allergies automatically puts the baby at high risk of developing an allergy. Later on in the book, we'll also cover the not-so-common allergies seen in children, such as drug reactions and insect bites (see Chapter 8).

Eczema

Eczema (also known as atopic dermatitis) refers to dry, sensitive, inflamed, itchy skin. It's very common and known to affect around 15–30 percent of children and 2–10 percent of adults. People with eczema find it hard to keep the moisture in their skin, so it becomes dry and easily irritated. This causes chemicals to be released, which worsens the irritation and makes them want to scratch. But scratching only makes the skin itchier, so the cycle repeats itself. It's as frustrating as it sounds.

A child with eczema has sensitive skin, which is irritated very easily. Their sensitive skin is often itchy, and scratching or rubbing the skin results in the eczema rash. In fact, you might say that eczema is "the itch that rashes."

Why does my child have eczema?

Eczema is believed to be a genetic disorder resulting in sensitive skin. It tends to be associated with the predisposition to become allergic to foods and substances in the air, such as pollens, molds, animal dander and dust mites. Some children with eczema later develop severe allergic reactions to foods, and many develop asthma and hay-fever symptoms as they get older. Often, there is someone else in the family with eczema, asthma or hay fever, but this is not always the case. There are many external factors that may influence eczema on a day-to-day basis, such as irritants (for example, soaps and household cleaners).

How is it diagnosed?

Diagnosing eczema should only be done by a qualified medical professional based upon patient history and physical examination. There are some features that on physical examination may strongly suggest eczema. For example, in infants and children, the eczema is often found on the scalp, knees, elbows and cheeks.

How is it managed?

While there is no cure, eczema is quite treatable. Here are some ways to manage your child's eczema:

- Protect your child's skin by applying moisturizer every day.
- Treat flare-ups by using ointments or creams prescribed by your doctor.
- Control itching by using antihistamines and a cold compress for the affected area.

Handwashing and eczema

Hand hygiene is extremely important, especially during flu season and viral pandemics. Alcohol-based hand sanitizers are frequently favored for their virus-killing skills, but they are very harsh on the skin of children with eczema. Proper handwashing with soap is better than using hand sanitizer, especially if you're able to apply moisturizer to your child's hands immediately after washing. Carry a small tube or jar of moisturizer wherever you go, so it is on hand to regularly apply to your child's skin.

Will my child outgrow their eczema?

The good news is that in most cases your child's eczema will gradually improve as they get older. Many kids are significantly better by the age of three years, and most will have only occasional trouble by the time they are teenagers. It is estimated that about two-thirds of children outgrow their eczema, although they may always have a tendency for dry skin. Only a few continue to have problematic eczema in adult life.

Asthma

Most of us are familiar with asthma. It typically has symptoms such as wheezing, shortness of breath and chest tightness. It's a disease characterized by recurrent attacks of this breathlessness and wheezing, which vary in severity and frequency from person to person. Symptoms may occur several times a day or week in those affected, and for some people symptoms become worse during physical activity or at night. During an asthma attack, the lining of the bronchial tubes swells, causing the airways to narrow and reducing the flow of air into and out of the lungs. Recurrent asthma symptoms frequently cause sleeplessness, daytime fatigue and reduced activity levels, and they can result in substantial time off school or work. The positive news from a large Australian study is that almost two-thirds of children with mild, intermittent asthma symptoms did outgrow their asthma and had no symptoms during adulthood.

However, those with persistent or more severe asthma in childhood or those who also have hay fever are less likely to outgrow their asthma, with their symptoms tending to persist into adulthood. Asthma can have severe complications and should be treated very seriously.

What are the triggers?

The fundamental causes of asthma are not completely understood. The strongest risk factor for developing asthma is the combination of genetic predisposition with environmental exposure to inhaled substances and particles that may provoke allergic reactions or irritate the airways, such as:

- indoor allergens (for example, house dust mites in bedding, carpets and stuffed furniture; pollution and pet dander)
- outdoor allergens (such as pollens and molds)
- tobacco smoke
- air pollution

Other triggers can include cold air, extreme emotion (such as anger or fear) and physical exercise. Even certain medications can trigger asthma, such as aspirin and other non-steroid anti-inflammatory drugs, and beta-blockers (which are used to treat high blood pressure, heart conditions and migraine).

How is it treated?

Although asthma cannot be cured, appropriate management can control the disease and enable children to enjoy a good quality of life. Short-term medications are used to relieve symptoms. Medications such as inhaled corticosteroids are needed to control the progression of severe asthma and reduce the likelihood of more severe complications. On occasion, asthma can be extremely severe and lead to children being admitted to hospital. If not properly controlled, an extremely severe asthma attack can be fatal.

Children with persistent symptoms must take long-term medication daily to control the underlying inflammation and prevent symptoms and flare-ups. But medication is not the only way to control asthma. It is also important to avoid asthma triggers. Over time, you and your child will learn what triggers your child's asthma, and what they need to avoid.

Swimming is often promoted as an exercise that can help kids overcome their asthma. While swimming can help improve their fitness, there is no evidence that it reduces the occurrence of kids' asthma attacks. Excess chlorine in swimming pools is harmful to the lungs of asthmatics, so kids with asthma should avoid swimming pools with a very strong chlorine smell or containing water that irritates the skin and eyes.

If your child has asthma, they should have a clearly written asthma action plan from their doctor or nurse that documents which medications to take regularly and which medications to take as needed, as well as when to seek a medical review. This plan should be reviewed every six months. Their school (or preschool) should also have an asthma first-aid plan for your child. Everyone who cares for your child should be aware of their medication and first-aid plans.

Hay fever (allergic rhinoconjunctivitis)

When we use the common term "hay fever," we're actually referring to allergic rhinoconjunctivitis. Hay fever symptoms include a blocked or congested nose, watery discharge from the nose, sneezing and nose itch, as well as watery eyes, itching around the eyes and eye redness.

The most frequently encountered allergens involved in children's hay fever include indoor allergens that are present throughout the year (such as house dust mites and pet allergens) and seasonal outdoor allergens (such as pollen from grass and trees). Food allergens do not cause hay fever.

How is it diagnosed?

Symptoms of hay fever are non-specific—they also occur during common viral nose infections. However, the common cold should last less than ten days. So children with nasal symptoms of a cold that last longer than ten days, or those who seem to have colds all the time, probably have hay fever.

The diagnosis of hay fever in children can be made by taking a history. Children are often seen for a cough and the general feeling of being unwell, and hay fever is only recognized when symptoms such as nasal congestion, discharge and itch are specifically noted. Children with the condition have twice the likelihood of a history of asthma, eczema, chronic sinusitis and otitis media with effusion (glue ear) than healthy children. It's useful for doctors to screen patients for airborne allergens, such as pollen, so that parents and children with hay-fever allergy can avoid the particular allergen. Skin-prick tests or the measurement of specific IgE (allergy) antibodies in blood are equally good options. Screening should include house dust mites, grass and tree pollen, and pet allergens. Screening for food allergens is not necessary.

How is it managed?

There are several ways to manage hay fever: educate the child, their parents and caregivers on hay fever; avoid allergens and tobacco smoke; and take pharmaceutical drugs.

Interestingly, there's not much data on the effect of allergen avoidance for children with hay fever. We know that complete avoidance of allergens is extremely effective, as children with seasonal hay fever and pollen allergy are symptom-free when outside of pollen seasons. But it isn't clear just how much allergen exposure needs to be reduced to improve symptoms.

For children *with asthma* who are sensitized to house dust mites, avoiding mite allergen is effective (according to evidence from several randomized controlled trials). However, studies found no benefit of mite avoidance in children without asthma. Additionally, there is no strong evidence that avoiding pet allergens in children with hay fever is beneficial.

Can my child take medication?

Intranasal corticosteroids are the first choice for the maintenance of persistent or severe hay fever, and have been found to be the most effective treatment of hay fever in adults. They reduce nose and eye symptoms and improve overall daily functioning. They are more effective than antihistamines (according to evidence from a systematic review of randomized controlled trials).

While there has been no strong evidence for intranasal corticosteroids in children, their benefits are thought to apply to children. Kids might find it upsetting or uncomfortable to have the hard nozzle of the nasal spray inserted into their nose, so it is a good idea to first apply Vaseline to the nostril to help the nozzle of the nasal spray slide in gently. Fortunately, the normal daily doses of most intranasal corticosteroids seem to have no side effects in children.

Some antihistamines can be effective and safe for treating hay fever in children. They reduce the feeling of being unwell and even improve learning ability (according to evidence from a randomized controlled trial). Intranasal decongestants relieve nasal congestion for a short period of time, but they do not improve nasal itching, sneezing or a runny nose. Prolonged use for more than ten days is not recommended, because this can lead to the decongestants irritating the nasal lining rather than making it better.

Will my child outgrow hay fever?

Children may outgrow the allergy as their immune system becomes less sensitive to the allergen. However, only about

10–20 percent of children will experience a complete elimination of symptoms with time (in other words, truly outgrow the hay fever). Around half of affected kids find that their symptoms get better with time; for the other half, things tend to stay around the same. The good news is that most cases of hay fever can be effectively managed.

CHAPTER 7
FOOD ALLERGIES

THE GUT AND FOOD ALLERGIES

Gut bacteria are clearly important for allergies in general. The presence of certain bacteria seems to play a role in children outgrowing their allergies.

A multicenter study in the United States examined 226 young children with confirmed cow's milk allergy. These children were followed until they were eight years old, and it was found that the cow's milk allergy had resolved in almost 60 percent of them by that time. The study found that gut microbiota composition at three to six months of age was associated with cow's milk allergy resolving by eight years of age—those who ultimately outgrew their milk allergy were more likely to have had a particular group of bacteria in their gut.

The multicenter study also found that the kids who outgrew their milk allergy had fewer dietary fats processed by bacteria. Processed fats are known to make it easier for allergens to cross the intestinal barrier. These same fats in cow's milk are also known to favor Th2 (allergy) protein production. It is therefore possible that the reduction in bacteria-processed fats from their diet may be one important part of kids outgrowing their milk allergy.

Some very exciting research was published in January 2019. This work was carried out to understand how our commensal gut bacteria (those that live within us without hurting us) can affect the food–allergy response. The team took gut microbes from healthy babies and from babies with cow's milk allergy, and colonized the gut of germ-free mice. Those that received microbes from infants *without* cow's milk allergy didn't have a reaction when exposed to cow's milk. Microbes from allergic infants *did* trigger anaphylaxis when the mice were exposed to cow's milk. This proved that the gut microbiota is pivotal to determining if someone develops an allergic reaction to food. Research continues on practical treatments involving microbes and allergies, but we're making good progress. (See Chapter 12 for more information on developing research.)

FOOD ALLERGIES FOR BEGINNERS

While research into gut microbes and food allergies is ongoing, there are some practical things that parents should know about food allergies. There is a lot of anxiety out there about food allergies, and it is not unheard of for parents to give their babies peanuts and other allergenic foods for the first time while parked in cars directly outside a hospital.

What's the difference between food allergy and food intolerance?

An allergic reaction to a food is an immune response. When a person has an allergic response, it is the body incorrectly identifying something harmless as being harmful and launching an attack as if it were fighting an infection.

A food intolerance is quite different, and occurs when a specific food has a *metabolic* response where the body cannot digest a particular food. It is not related to the immune system, like an allergy is, and it cannot cause anaphylaxis. Lactose and gluten intolerances are two popular examples of conditions where a person cannot process a food.

Food labels

As a parent, you can learn to read food labels carefully and be aware of the ingredients that trigger allergies. You can also teach your child and other family members to be aware of allergenic ingredients.

Known food allergens must be declared in plain language on packaging and are usually highlighted in bold text after the ingredients section in a "Contains" statement. The precautionary statement "May contain" is sometimes seen in labeling after the "Contains" statement. It refers to possible cross contamination with allergen traces, for example by workers in a food manufacturing plant using the same gloves to handle different foods. It's generally advised to avoid a food if an ingredient you're allergic to is listed in a precautionary label.

Here is a sample ingredient list, with allergens highlighted in bold text at the end:

Ingredients: Wheat flour, vegetable oil, sugar, cornflour (wheat), hazelnuts, salt, milk solids, sesame, raising agents (E330, E341, E500), yeast, emulsifier (E322, soy), egg.
Contains: Wheat, tree nuts, milk, sesame, soy, egg.
May contain: Peanuts.

In the United States, the *Food Allergen Labeling and Consumer Protection Act* (FALCPA) became law in 2004. The FALCPA requires foods to be labeled to identify eight major food allergens: milk, egg, fish, crustacean shellfish, tree nuts, wheat, peanuts and soybeans. In 2021, the FASTER Act passed, requiring sesame to be added to the list of allergens on US food labels starting in 2023. In Canada, along with the eight major food allergens, sesame, triticale (a hybrid of wheat and rye) and mustard seeds are considered priority allergens that need to be labeled on food packaging.

How will I know if my baby has a food allergy?

The best way to find out if your baby has a food allergy is to be aware of the foods that trigger allergic reactions, introduce just one allergy-inducing food at a time, and carefully monitor your child for signs of a reaction when they first consume these foods.

Tips for testing a common allergy-inducing food

- You should always introduce a new allergen at a set time, either in the morning or at lunchtime, as this will allow you to monitor the reaction throughout the afternoon.
- Pick a quiet day, when you will be at home together or close to home to monitor your baby for a reaction.
- If you find that you need to sleep or be away from your baby for some reason in the hours after they try a common allergy-inducing food, make sure that you get a trusted person to watch for signs of a reaction. Instruct that person

to wake you or get in contact immediately should a reaction develop.

- Babies should not be unmonitored after taking a food allergen, as a reaction may take hours to develop.
- If there is no reaction after your baby first eats the allergenic food, repeat the same allergen for three days in a row and continue to look for a reaction.
- If there is no reaction after three days, then it is safe to assume that your baby is not allergic.

Symptoms of a reaction to allergy-inducing foods

Mild to moderate reactions can include:

- a rash
- an itchy nose
- a runny nose
- hives or welts
- swelling of the lips/face/eyes
- tingling of the mouth
- abdominal pain
- vomiting
- diarrhea

But any of the following signs can be indicative of a severe allergic reaction, so if you notice them, seek immediate medical attention or call for an ambulance:

- difficulty/noisy breathing
- swelling of the tongue
- swelling or tightness in the throat

- difficulty talking and/or hoarse voice
- wheeze or persistent cough
- persistent dizziness
- your child collapsing, or becoming pale and floppy

If a serious allergic reaction is confirmed, your baby will be referred to a pediatric allergist. In the meantime, you need to make sure that they have zero contact with the food that caused the allergic reaction. Your pediatric allergist may carry out a number of tests on your baby and develop an allergy action plan for you and your baby. In some cases, your growing child will need to avoid that particular food for life; you will be given information about this as well as instructions on how to use an EpiPen Junior or EpiPen. Giving your child a medical-alert bracelet and ensuring that they always wear it will help others to easily identify that your child has a confirmed serious allergy.

Tips on using an EpiPen Junior

For young children weighing between 15 and 30 kilograms (33 to 66 pounds), an EpiPen Junior is recommended. For children who weigh less than 15 kilograms (33 pounds), your pediatric allergist will help work out the correct dosing of adrenaline, and this will form part of your Anaphylaxis Emergency Plan.

The EpiPen Junior sits inside a pen-shaped carry case with a green top. The pen should be gripped away from the orange tip, as this contains the needle. Once the orange tip is pointing down over a child's thigh, the blue safety release can be pulled off. An easy phrase to remember the correct orientation of the EpiPen Junior is "Blue to the sky, orange to the thigh."

Push the EpiPen Junior down hard into the thigh muscle until a click is heard or felt, and hold it in place for three seconds. The EpiPen Junior can then be removed. You should call emergency immediately after administering an EpiPen Junior to get urgent medical assistance.

Are some foods more likely to spur an allergic reaction than others?

Just nine foods are responsible for 90 percent of food allergies: milk, wheat, eggs, soy, peanuts, tree nuts, sesame, shellfish and fish. Berries, seeds, corn and other foods can also be problematic. At least half of the children who develop a food allergy during the first year of life outgrow it by the time they are two or three years old. Some reactions to food (for example, milk or soy) are more often outgrown than others. Some allergies fade as kids mature. Wheat, milk and egg allergies are most commonly left behind by growing kids, while shellfish, fish, tree nut and peanut allergies tend to be lifelong. While there is no cure for food allergies, scientists continue to study ways to lessen symptoms so that children can tolerate problematic foods.

THE NINE MAJOR FOOD ALLERGIES

Cow's milk allergy

Cow's milk allergy is one of the most common food allergies affecting young children in the developed world. It is estimated that around one in 50 infants in North America is allergic to cow's milk. The good news is that, just like an egg allergy, most children will outgrow it—around 85 percent of cow's milk allergy cases will disappear by the age of five.

Cow's milk allergy presents in two forms. It can be an IgE mediated (antibody) allergy or a non-IgE mediated allergy. IgE mediated symptoms occur rapidly, while non-IgE mediated symptoms can take hours or days to appear. A family history of allergy or eczema is very influential and puts the infant at significantly greater risk of developing cow's milk allergy. The steady beat of the atopic march can also give us some helpful clues. Early-onset eczema (appearing before three months of age) or having more severe eczema than usual is a sign that the infant has a high risk of developing cow's milk allergy.

Symptoms

Cow's milk allergy is often missed early on, because it has a fairly diverse range of symptoms:

- In the gut, it can lead to vomiting, regurgitation, diarrhea, bloody stools and/or abdominal pain.
- In the lungs it can lead to wheezing or a chronic cough; sometimes we see a runny nose.
- In the skin, we might see inflammation; swelling of the face, eyes or lips; and/or raised red bumps or hives.

You're probably thinking that these are very broad symptoms—how can we distinguish these from the "ordinary" symptoms that a baby might be experiencing? The timing is everything. Symptoms usually develop within the *first week* that cow's milk is introduced to an infant's diet. So it is really important to be alert to any of these possible symptoms shortly after your baby starts having cow's milk. If you have any concerns, then please see your local doctor for guidance. They may recommend a period of cow's milk elimination followed by re-exposure (also known as rechallenge). Recurring symptoms are seen as strong evidence of a cow's milk allergy.

The timing of when symptoms occur can give us an idea of what kind of immune reaction might be taking place to cause the cow's milk allergy.

IgE mediated allergy

When your child has an IgE mediated allergic reaction, IgE antibodies trigger a cascade of allergic reaction symptoms. Blood tests will show raised levels of IgE antibodies, and skin-prick testing will cause large welts. The time from eating to reaction is very short and almost always less than one hour. Typically, the reaction will occur within 30 minutes of eating, and symptoms can be prolonged. In the case of IgE mediated cow's milk allergy, just a small amount of milk can trigger a reaction.

Non-IgE mediated allergy

On the other hand, non-IgE mediated food allergies are caused by a reaction involving other components of the immune system apart from IgE antibodies. Non-IgE mediated cow's milk allergy results in delayed inflammation. Diagnostic tests, such as blood tests and skin-prick testing, are usually negative. The time from eating to reaction is delayed and almost always at least two hours. In fact, a response can occur more than a day after exposure to cow's milk, which can make it a little difficult to track. We also know that, unlike IgE mediated cow's milk allergy, a large volume of milk is required to trigger a reaction. The symptoms are generally more non-specific than you'll find in a IgE mediated cow's milk allergy. Symptoms can include delayed vomiting and diarrhea, loose and frequent bowel movements, blood and mucus in the stools, general irritability and delayed eczema. Children with non-IgE mediated cow's milk allergy should see an experienced medical practitioner.

IgE and non-IgE mediated cow's milk allergy almost always occurs in infants and young children. It is very uncommon to see a cow's milk allergy develop in an older child.

LACTOSE INTOLERANCE

I should point out that we must not confuse cow's milk allergy with lactose intolerance, which is NOT an allergy. It's best to think of lactose intolerance as an inability to properly digest lactose—it does not result in damage to the child's gut. If untreated, cow's milk allergy can certainly lead to damage to the gut. Lactose intolerance also usually presents a fair bit later in early childhood than cow's milk allergy, and may not be a real issue until three or four years of age.

Management

All children with a cow's milk allergy should see an experienced medical practitioner for ongoing assessment until they've outgrown their allergy. If their reaction is very severe (anaphylactic), then a specialist-led plan will need to be formulated with the parents and will usually include access to an EpiPen Junior or an EpiPen.

Surprisingly, cow's milk allergy can occur in exclusively breastfed infants, which means it is possible that a reaction to cow's milk is passed to the baby from the mother's diet. We mentioned in Chapter 3 how an unborn baby can become sensitized to food allergens in the womb. However, breastfeeding is safe from all forms of dairy allergy because human milk is quite different from cow's milk. It's important to have a discussion with your doctor about whether or not to eliminate cow's milk from your diet when breastfeeding. In general, cow's milk protein should be eliminated from the mother's diet if the breastfeeding baby has an immediate reaction, such as anaphylaxis. But in many delayed non-IgE cases, mom's intake of cow's milk is usually fine for the baby and rarely needs to be restricted.

If the baby is not exclusively breastfed and has cow's milk allergy symptoms, the first thing to do is to remove cow's milk from their diet. This means strictly no cow's milk–based formulas, including A2, partially hydrolyzed, anti-reflux or lactose-free milk. Other animal milks, such as goat milk, should also be avoided. Soy-based formula is also not recommended for infants under six months of age. This is because infants under six months who have a cow's milk allergy have a much higher rate of reacting to the presence of soy than infants older than six months.

The best formula choice for infants under six months who have cow's milk allergy is an extensively hydrolyzed formula, which can be continued on from that age. An extensively hydrolyzed formula has been greatly modified from cow's milk by enzymes that break down most of the cow's milk proteins. If your baby reacts to or does not tolerate extensively hydrolyzed formula, an amino acid–based formula can be used. This is necessary in around one in ten infants with a cow's milk protein allergy. Amino acid–based formulas are 100 percent amino acids and don't contain any cow's milk protein. These specialized infant formulas, particularly amino acid–based formulas, can be quite expensive, and in the United States you should check with your health insurance provider to see if your coverage includes them. The majority of the time, health insurance will cover this cost for a medically confirmed cow's milk allergy. In Canada the cost of these specialized formulas can often be substantially covered by your province under a Special Diet Allowance; you can liaise with your child's pediatrician or dietitian to find out more about this. Getting an appropriate assessment from a pediatric immunologist is important, because long-term use of a rice protein–based extensively hydrolyzed formula may not be suitable in some rare non-IgE immune conditions.

I want to stress that there is no convincing evidence to support the use of extensively hydrolyzed formula to prevent allergy in infants or children—breastmilk should still be the first port of call, if possible. This is based upon a good systematic review published in the *British Medical Journal* in 2016. Hydrolyzed formula is recommended for babies when there is confirmed cow's milk protein allergy, but it is not used to prevent it, even if there is a high risk of the baby developing an allergy.

Finally, it is important to read the labels of solid foods to ensure that you choose foods for your infant that do not contain cow's milk protein or any other animal milk. In Australia, by law food manufacturers must indicate on the label if a product contains cow's milk. Some examples of ingredients that include cow's milk are: milk, non-fat milk solids, milk solids, milk powder, condensed milk, evaporated milk, whey, casein, lactalbumin, lactoacidophilus, margarine, butter, butter oil, buttermilk, yoghurt, cream, custard, ice cream, cheese, chocolate, butter fat, curds and lactose. Where there is any doubt, I'd recommend consulting an accredited practicing dietitian for expert advice.

WHAT YOU CAN DO AS A PARENT

- For non-exclusively breastfed babies who are under six months of age, extensively hydrolyzed formula is recommended as the first choice for cow's milk allergy. It can be continued past six months of age.
- If a baby reacts to or doesn't tolerate extensively hydrolyzed formula, an amino acid based formula can be used. It can be continued past six months of age.
- In the United States, ask your health provider if your child can be covered for the cost of these specialized infant formulas; in Canada, check with your child's pediatrician/dietitian about applying for your province's Special Diet Allowance.
- For infants over six months of age, soy formula can be used for cow's milk allergy.

- There is no evidence to support the use of extensively hydrolyzed formula to prevent allergy in infants or children.

Egg allergy

Egg allergy is common in North America, with an estimated 2 percent of children having an allergy to raw egg. It is usually an IgE antibody-driven reaction, but occasionally children can present with a non-IgE egg allergy.

There is good evidence that giving your baby cooked egg at around six months of age prevents the development of an egg allergy. Even if your baby is considered low risk for an egg allergy, it is still a very good idea to give them egg in their first year.

Symptoms

An egg-allergy reaction will cause similar symptoms to other food allergies such as cow's milk. The reactions include:

- In the gut, it can lead to vomiting, regurgitation, diarrhea, bloody stools and/or abdominal pain.
- In the lungs it can lead to wheezing or a chronic cough; sometimes we see a runny nose.
- In the skin, we may see inflammation; swelling of the face, eyes or lips; and/or raised red bumps or hives.

The timing of the reaction is again very important. As egg allergy is usually an IgE allergy, which means that symptoms occur fairly quickly after eating—as in my daughter Olivia's case. Knowing how much egg was eaten as well as the form of the egg (for example, raw, cooked or baked) is relevant for managing the egg allergy.

Management

There is currently no treatment for egg allergy. Fortunately, 80 percent of children will outgrow their allergy in time. Avoiding egg is the most appropriate treatment recommendation. It is the egg-white proteins that most children react to, especially the dominant protein, ovalbumin; only a few children experience a reaction to egg yolk. As it's quite difficult to entirely separate an egg white from the yolk, avoiding the entire egg is recommended.

Routine vaccines such as the measles, mumps and rubella (MMR) and influenza vaccines are perfectly fine to administer to children with egg allergy. Reactions can occur with the Q fever and yellow-fever vaccines, which are recommended when traveling to particular countries. Seek the advice of a pediatric immunologist if your baby needs these vaccines.

Egg allergy resolves in stages, starting with tolerance to well-cooked egg (such as you'd find in a cake) and then lightly cooked egg (such as scrambled eggs). The last stage of tolerance is to raw egg, but children and adults rarely consume raw egg—and young children shouldn't be given raw egg because of the risk of food-borne illnesses, such as salmonellosis.

For Olivia, given that she had a significant reaction, her immunologist recommended that she avoid eggs for three months. When she was nine months old, we gave her a small mouthful (around a quarter of a teaspoon) of sponge cake, which she tolerated with no reaction. Over the next few weeks and months, we built up her consumption of cooked egg with cakes, biscuits and egg noodles. We then introduced lightly cooked egg in the form of a quarter of a teaspoon of scrambled egg, which she also tolerated with no reaction. Since then, she has been fine with all food containing egg. If Olivia had developed a significant reaction to the scrambled-egg trial, we would have gone back to giving her well-cooked egg for at least another three months before trying scrambled egg again.

WHAT YOU CAN DO AS A PARENT

- If your child has an egg allergy, avoiding egg is key. It is recommended that the child not consume any part of the egg (yolk or white).
- Check the ingredients list on packaged foods for the presence of egg.
- The introduction of well-cooked egg to the diet at around six months of age helps to prevent the development of egg allergy for any child.
- Egg allergies resolve in stages. Introduce well-cooked egg and then lightly cooked egg. Raw egg should not be trialed in young children due to the risk of infections.

Peanut allergy

One in 50 children and one in 200 adults have a peanut allergy. Unlike egg and cow's milk allergies, we know that a peanut allergy is less likely to be outgrown and therefore tends to be a lifelong condition. Infants with severe eczema or an egg allergy have a high risk of developing a peanut allergy. Small traces of peanuts can cause anaphylaxis in a highly sensitized child. In North America, peanuts are the food most likely to cause death from anaphylaxis in children.

Not so long ago peanut was not a common cause of anaphylaxis in Asia, but this is rapidly changing with the adoption of a Westernized lifestyle. A 1999 Singaporean study did not identify peanut or tree nuts as triggers of anaphylaxis, but a 2013 study on Singaporean children found that peanut had emerged as the most common food trigger of anaphylaxis.

Peanut allergies present most commonly during infancy and early childhood. Usually, a reaction occurs the very first time an infant or toddler is given a bit of bread with peanut butter. Children can become sensitized to peanuts in a number of ways, with the most common believed to be from peanut protein present in breastmilk. So why aren't we advocating to restrict peanuts in the mother's diet while she is breastfeeding? Unfortunately there is no good evidence that doing so is actually beneficial in preventing the allergy in kids.

Nor is there any good evidence to suggest that restricting peanuts from the mother's diet while she is pregnant lessens the risk of childhood peanut allergy. On the contrary, in Chapter 3 we referred to the GUTS2 study, which showed that mothers who consumed *more* peanuts during pregnancy had a lower incidence of peanut allergy in their kids. The current recommendation is for no maternal dietary restrictions during pregnancy or breastfeeding. If confirmed peanut allergy does occur in a baby while they are still being breastfed, then restricting peanuts from the mother's diet is, of course, sensible and appropriate.

Children can also be exposed to the peanut allergen when they are kissed. The first time that a baby is kissed on the cheek by someone who has just eaten peanut butter can lead to a bright red welt. It only takes 1/2000th of a single peanut to lead to a reaction in some highly sensitized people, so if they are kissed by someone who has just eaten peanuts, it can actually lead to anaphylaxis. Brushing your teeth, rinsing your mouth or chewing gum after a meal that includes peanuts won't help.

The problem that even such small traces of peanut can cause a reaction is made more challenging by the fact that repeated exposure to peanuts heightens recognition of this allergen by the immune system and strengthens the allergic response. Every time a child is re-exposed, less and less of the allergen is required to cause a reaction. Inhaling food particles containing peanut allergen can eventually be enough to trigger a reaction.

Symptoms

Symptoms of an allergic reaction to peanuts include:

- a runny nose
- tightening in the throat
- shortness of breath or wheezing
- a skin reaction, such as hives or redness
- tingling or itching in the mouth or throat
- diarrhea, nausea, stomach cramps and/or vomiting

Management

Twenty percent of children outgrow their peanut allergy, but this means that 80 percent will have a persistent allergy that generally lasts their whole life. The best treatment for those who have peanut allergy is to completely avoid peanuts. Food labels should make note of the presence of peanuts, and parents should pay particular attention to what is in manufactured foods. For children, exposure to peanut butter is more likely to be problematic, given that other children who have eaten it can easily contaminate areas such as playground equipment. For this reason, day care centers that have a child with known peanut allergy will advise parents not to bring any peanut butter or nuts to the premises. Homes with a child who has known peanut allergy should do the same, as contamination of surfaces from peanut butter or nuts can easily occur.

For families with a child who has peanut allergy, eating out at restaurants or buying takeaway can be major risks, as the food may not be carefully labeled for potential allergens. The consumption of new or exotic foods with many unknown ingredients can be very dangerous, so it is best to avoid this exposure completely.

Vitamin D and ultraviolet light

In Australia, the rate of peanut allergy increases the farther south children live. A similar phenomenon has been observed in North America, with one study finding that there were eight to twelve prescriptions of EpiPen (used to treat anaphylaxis) for every 1000 people in New England compared to only three prescriptions in the southern US states. The finding that children are more likely to develop allergies the farther they live from the equator may be explained by their not receiving enough vitamin D from sun exposure. The HealthNuts study found that children who were vitamin D deficient were eleven times more likely to have peanut allergy. It is certainly possible that low levels of vitamin D can skew the immune system towards activation rather than tolerance.

However, more recent research from The University of Western Australia has found that it may be ultraviolet light rather than vitamin D that is helpful in preventing allergies in these children. In their study, newborns received either a vitamin D supplement or a placebo until six months of age. Some of these infants also wore a device to measure their direct ultraviolet-light exposure. At three and six months of age, the babies' vitamin D levels were measured. The presence of eczema was assessed at six months of age. The study showed that greater direct ultraviolet-light exposure in early infancy was associated with a lower incidence of eczema by six months of age. Vitamin D supplements had no impact on the rate of eczema. It isn't yet known if ultraviolet-light exposure can reduce the rate of peanut allergies, but clinical trials have been planned.

Ultraviolet-light exposure for peanut-allergy prevention was assessed in a different way by researchers from the University of Florida. They showed that when pulsed light (which contains mostly ultraviolet light) is applied to whole peanuts, it can reduce 80 percent of allergens. It has previously been tested on peanut-protein extracts to remove 90 percent of allergens.

Parents should encourage their babies to enjoy some natural sunlight, but it's very important to follow sensible safety precautions when out in the sun. The Centers for Disease Control and Prevention has an excellent website (cdc.gov/cancer/skin/basic_info/children.htm) which provides very useful information on sun protection.

The best times to take your baby outside are during the early morning or late afternoon; in the middle of the day, UV levels are at their highest. Widespread use of sunscreen on babies under six months of age is generally not advised, as it can be quite irritating to their skin. It's better to use shade, clothing and broad-brimmed hats to protect them from the sun.

Feed peanuts to kids early

There is now good evidence that regular peanut intake before twelve months of age can reduce the risk of developing a peanut allergy. Over a decade ago, the thinking was very different—scientists believed that children at high risk of the allergy should avoid peanuts until they turned three, when they would have a stronger, more mature immune system. But this strategy consistently failed to prevent peanut allergies.

The 2015 UK Learning Early About Peanut Allergy (LEAP) study led the way on the new thinking about exposing babies to peanuts in their first twelve months of life. Worldwide, no cases have been reported of anaphylaxis death caused by peanuts in the first twelve months of life. This suggests that the first twelve months is the ideal time to expose a baby to peanuts and hopefully create immune tolerance.

The single-center LEAP study included 640 babies aged between four and eleven months of age with severe eczema, egg allergy or both. The babies either consumed or avoided peanut-containing foods until they were five years old. Babies with a confirmed peanut allergy were excluded from the study. Peanut-based skin-prick tests were administered to the participants, and the babies were divided into positively and negatively sensitized based on the skin-prick results. Positively sensitized babies are very likely to develop peanut allergy in childhood, whereas negatively sensitized babies are less likely. At five years of age, all the children were given peanuts as a food challenge.

The LEAP study found that eating peanuts regularly, beginning in the first eleven months of life, resulted in a significantly *smaller* proportion of children with a peanut allergy at the age of five compared to avoiding peanuts. Consuming peanuts was associated with an 86 percent reduction in the prevalence of peanut allergies at five years of age among children who had had a negative skin-prick test when they were babies. For those children who had a positive skin-prick test when they were babies, meaning they were more at risk of developing an allergy, there was a 70 percent reduction in the prevalence of peanut allergy.

An extension to the LEAP study, the LEAP-ON study was designed to see if the children who had consumed peanuts in the LEAP study would remain protected against peanut allergies after they stopped eating peanuts for twelve months. The researchers observed that there was no significant increase in the prevalence of peanut allergies. The findings showed that, even after not eating any peanuts for a full year, nearly all the children in the study, who were at high risk of developing peanut allergy and had consumed peanuts regularly up until the age of five, continued to be tolerant.

What LEAP and LEAP-ON have found is that the early introduction of peanuts to babies at high risk of developing an allergy dramatically reduces rates of peanut allergies at five years of age, and this effect continues after a year of not eating peanuts. It is highly likely that a state of sustained unresponsiveness to peanuts has continued in those children. If this effect remains steady years into the future, then we can say that true immune tolerance has occurred.

How to safely introduce peanut butter to your child

The Australasian Society of Clinical Immunology and Allergy (ASCIA) has brilliant resources for parents on how to introduce solid allergenic foods to their children. The following step-by-step guide to introducing peanut butter to your child for the first time is adapted from the ASCIA website (allergy.org.au). If you'd like other recipe options for preparing peanut butter, you can check out Appendix D of the Prevention of Peanut Allergy Addendum Guidelines by the US National Institute of Allergy and Infectious Diseases (NIAID), which was released in early 2017 (niaid.nih.gov/sites/default/files/addendum-peanut-allergy-prevention-guidelines.pdf).

1. Using the tip of your index finger, put a small dab of smooth peanut butter or paste on the inside of your baby's bottom lip (right in the middle). Try to avoid getting any on their skin.

2. Watch for an allergic reaction for a few minutes. If nothing happens, feed your baby a quarter of a teaspoon of smooth peanut butter or paste, and keep a close eye on them for the next half hour for any sign of an allergic reaction.

3. If there's no reaction after thirty minutes, you can give your baby a little bit more. Try half a teaspoon of smooth peanut butter or paste, then watch them closely for another half hour for any sign of an allergic reaction.

4. If everything seems okay, you can introduce half a teaspoon of smooth peanut butter or paste into your baby's diet once a day for the next three days, watching them each time for any sign of a reaction.

5. You should continue to include smooth peanut butter or paste in your baby's diet at least three times a week. If at any stage if you suspect an allergic reaction, stop feeding your baby peanut products immediately and seek medical advice.

WHAT YOU CAN DO AS A PARENT

- Prevention is the best medicine: avoiding contact with peanuts altogether is the best possible treatment for children with a peanut allergy.
- Babies should have some natural sunlight, but be sure to follow safety precautions for sun protection.

- There is good evidence that regularly eating peanut products before twelve months of age can reduce the risk of a future peanut allergy by developing immune tolerance.

Tree-nut allergy

The tree-nut family includes chestnuts, pistachios, walnuts, almonds, cashews, hazelnuts, hickory nuts, brazil nuts, macadamias, pecans, pine nuts, and others. Tree nuts are actually different from peanuts, because they come from a different plant family.

Symptoms

The symptoms of a tree-nut allergy are very similar to a peanut allergy:

- abdominal pain, cramps, nausea and/or vomiting
- diarrhea
- difficulty swallowing
- itching of the mouth, throat, eyes, skin and/or any other area
- nasal congestion or a runny nose
- shortness of breath
- anaphylaxis (less common)

Management

Like peanut allergies, the cornerstone of management is successful avoidance of tree nuts, and prompt treatment of any allergic reaction should it occur. The following foods should be avoided by your child unless the chef who prepares the food can guarantee that there are no peanuts or tree nuts present: all Chinese,

Vietnamese and Thai dishes; baked goods (for example, biscuits and cakes); sauces (such as chilli sauces); and toppings, gravies and marinades. Some salads may also contain tree nuts, so check them carefully. When there is any doubt, avoid the food in question. Injectable adrenaline (EpiPen) is the treatment for anaphylaxis, and it must be administered promptly.

NUTS AND SEEDS

If a child has an allergy to a particular nut or seed, this doesn't necessarily mean that they will have a similar reaction to another type of nut or seed. Tree nuts such as almonds and cashews have different proteins to peanuts, and therefore someone who has peanut allergy might not react to tree nuts. It is impossible to predict a reaction to a nut or seed based upon a previous allergic reaction to peanuts, so proper allergy testing should be carried out.

Shellfish allergy

Shellfish allergies are one of the most common and serious food allergies worldwide, especially in Asia. This is largely due to the high rate of shellfish consumption in Asian countries. Shellfish is the leading cause of anaphylaxis in adults and older children in Hong Kong, Taiwan, Singapore and Thailand. The most commonly reported shellfish allergies, in decreasing frequency, are to prawn (shrimp), crab, lobster, clam, oyster and mussel.

Symptoms

Classic symptoms of an allergic reaction to shellfish include hives and facial swelling within minutes of exposure, which may proceed to anaphylaxis. Symptoms are fairly variable between children, and they can range from mild rashes to life-threatening anaphylactic shock. Thankfully, babies typically have much milder reactions.

Management

The best way to test whether or not your child has shellfish allergy is a carefully controlled oral-food challenge organized by your allergist. Clinical diagnostic tests can help to identify children who are likely to be able to tolerate a food challenge and those who are likely to have a bad reaction.

The most important component of the management of shellfish allergies is the complete avoidance of shellfish and other crustaceans. If your child has a history of suspected or proven reaction to shellfish, they and/or their caregivers should carry an EpiPen Junior or an EpiPen with them and know how to use it properly.

Fish allergy

Fish allergies affect about 0.5 percent of the population. They tend to be lifelong and are more common in adults than children. Kids who have a fish allergy are usually allergic to other types of seafood, too.

THE HEALTHY BABY GUT GUIDE

Those who are allergic to prawns can usually tolerate fish, but care must be taken because of the real risk of cross contamination in the storage, preparation and cooking of the seafood.

Symptoms

Mild symptoms of fish allergy include:

- itchy skin and rash
- hives
- swelling of the lips
- tingling of the throat and mouth
- runny nose
- chest tightness
- abdominal cramping
- nausea
- vomiting

These symptoms can progress to the more serious symptoms of anaphylaxis, including breathing difficulties. Touching seafood or even smelling fish, particularly cooked fish, can lead to the appearance of symptoms. Children who have known fish or shellfish allergy should not have that food cooked in their home because of the risk of the smell triggering a serious allergic reaction.

Management

Suspected fish allergies are confirmed with a skin-prick test or blood test. It's important to note that it isn't possible to test for all types of fish, as there is a limited range of commercially available skin-prick and blood tests. A careful history by a clinician and, in some cases, bringing samples of the suspected fish to a testing clinic can aid the diagnosis.

Similar to peanuts, tree nuts and shellfish, the best practice is the complete avoidance of fish, and prompt treatment of a serious allergic reaction with an EpiPen. Like all food allergies, it is important to inform anyone preparing food for your child that they have known fish allergy.

Soy allergy

A soy allergy occurs soon after exposure to soybeans or soy products. It is usually an IgE (allergy) antibody-driven reaction to soy proteins. IgE (allergy) antibodies can be detected using a skin-prick test or blood test.

Soy allergies are uncommon compared to peanut, egg or cow's milk allergies, and they are most often seen in young children with eczema (atopic dermatitis). Only 2–3 percent of young children will have a positive allergy test for soy. And of these positive children, less than 10 percent will have allergic symptoms when they actually eat soy—the other 90 percent won't have any symptoms, despite testing positive in the allergy test. For that small minority who do display symptoms from the soybean challenge, there is risk of anaphylaxis, so the challenge should only be undertaken with very close monitoring.

Only a small number of children with cow's milk allergies (about 15 percent) will also have or develop a soy allergy. There is no good evidence that eating soy products increases the risk of developing peanut allergy, even though both peanuts and soy are legumes.

Symptoms

Soy allergies in children often present as hives and/or vomiting. Sometimes a more severe allergic reaction (anaphylaxis) can occur. Rarely, delayed immune reactions to soy (which are not due to IgE allergy antibodies) can result in a flare-up of eczema or bowel symptoms in some children.

Management

People with soy allergy are advised to avoid foods that contain the following:

- soybeans/edamame
- soy flour
- soy milk and soy-milk products (for example, soy yoghurts, soy cheese, soy desserts, soy ice cream)
- soybean sprouts
- tofu (soybean curd)
- textured/hydrolyzed vegetable protein
- fermented products, such as miso (soybean paste), tempeh, soy sauce, tamari, bean curd and teriyaki

Children with soy allergy may outgrow their allergy—your child's immunologist should perform allergy tests every twelve months or so to determine if the allergy is still present.

Wheat allergy

Wheat grain is found in a whole variety of foods—cereals, pastas, crackers and even some hot dogs, sauces and ice cream. It is also found in non-food items, such as Play-Doh, cosmetics and bath products.

Wheat allergy symptoms will present soon after exposure to wheat products. It is usually an IgE (allergy) antibody reaction to wheat proteins, and is diagnosed through either a skin-prick test or blood test. Wheat allergy is most common in children; around 65 percent of children with wheat allergy will outgrow it by the time they are twelve.

Symptoms

The symptoms of a wheat allergy include:

- hives or skin rash
- nausea, stomach cramps, indigestion, vomiting and/or diarrhea
- stuffy or runny nose
- sneezing
- headaches
- difficulty breathing

If the symptoms are severe, anaphylaxis can occur.

Management

Managing a wheat allergy involves strict avoidance of wheat ingredients in both food and non-food products, as well as urgent treatment of any severe allergic reaction with an EpiPen.

Avoiding wheat can be more complicated than you'd think, as foods that don't contain wheat as an ingredient can be contaminated in the manufacturing process or during food preparation. As a result, people with wheat allergy should also avoid products that have wheat warnings on the label, such as "made on shared equipment with wheat," "packaged in a plant that also processes wheat" or similar statements.

The recent growth in gluten-free products is making it much easier to manage wheat allergy. Gluten is a protein found in wheat, barley and rye. Therefore, gluten-free foods should contain no wheat. Additional options for wheat-free foods include those made from other grains, such as corn, rice, quinoa, oats, rye and barley.

Sesame allergy

Sesame allergies are estimated to affect 0.1–1 percent of the overall population in the Western world, including approximately 0.6 percent of children. Only 20–30 percent of kids who have sesame allergy ever grow out of it, so most kids with a sesame allergy will have it for life.

Testing for a sesame allergy is potentially tricky. Oleosi, a major sesame allergen, is not detectable in skin-prick tests, though the other major sesame allergens are. Therefore, both skin-prick tests and blood tests for sesame allergy are highly recommended to avoid any false negative results.

There are differing amounts of allergens in the three varieties of sesame seeds. White sesame seeds contain the most allergens compared with brown or black seeds. However, all sesame seeds are allergenic, and sesame oil is considered highly allergenic, as is sesame flour.

Symptoms

The symptoms of sesame allergy include:

- hives or skin rash
- nausea, stomach cramps and/or vomiting
- itching in the mouth
- coughing
- difficulty breathing

Management

Strict avoidance of sesame seeds and their products is necessary. Common sources of sesame include: Asian dishes (especially Thai), hamburger buns, pizza crusts, breadsticks, rolls, bagels, pretzels, crackers, biscuits/cookies, candies, wafers, noodles, soups, ice cream, salad dressing, oils, tahini, hummus, margarine, cosmetics (such as lip balms, ointments, eye products and creams), pharmaceutical products (check labeling) and nutritional supplements.

As sesame is found so widely in different foods, it is important to avoid cross contamination. For example, if you have a child with a sesame allergy and you've just used a knife to spread hummus for another child's meal, then that knife is contaminated and you will need to use a new, clean knife when you're preparing food for your allergic child. Be aware of chopping boards, which can easily cross contaminate foods.

CHAPTER 8
OTHER ALLERGIES

ALLERGIC REACTIONS TO STINGS

The most common stinging insects that cause allergic reactions are bees (honeybees, bumblebees), vespids (yellow jackets, hornets, wasps) and fire ants. The last group—imported fire ants—are an important cause of anaphylaxis in the southern and southeastern United States, but not in Canada.

Levels of reaction to stings

An isolated local reaction occurs when the sting results in a rash or large local swelling alone. In such cases, there is less than 10 percent chance of developing a severe allergic reaction (anaphylaxis) with further stings.

A generalized reaction without life-threatening features can present as generalized hives (also known as urticaria), without difficulty breathing or a drop in blood pressure. This type of allergic reaction is uncomfortable but not dangerous. It is more common in children than adults and also has a less than 10 percent chance of progressing to anaphylaxis.

Anaphylaxis from insect stings in North America is usually due to allergic reactions to the venom of bees, yellow jackets, hornets, wasps or fire ants. Severe allergic reactions to insects are usually due to stings from bees, wasps or fire ants. Adults are at greater risk of anaphylaxis than children, but kids who have previously had anaphylaxis following a sting are at a high risk of another anaphylactic episode. Allergic reactions to stinging insects tend to persist, but the symptoms children experience are more likely to lessen over time compared to adults. Insect bites are a less common cause of anaphylaxis than insect stings.

Management

Usual first aid practices are adequate for treating minor allergic reactions to bees, which usually leave their barbed stinger in the skin and die. Flicking the stinger out from your child's skin as soon as possible will reduce the amount of venom injected. Use the edge of your fingernail or the edge of a hard plastic credit card. If possible, try not to squeeze the venom sac, as this may increase the amount of venom injected. By contrast, wasps and fire ants rarely leave their stinger in the skin.

Cold packs and soothing creams often help for minor reactions, and oral antihistamines can be useful for treating itching. Very strong and uncomfortable local reactions may need cortisone tablets to settle the swelling. A child with a history of anaphylaxis to an insect sting should be referred to an allergist.

Allergen immunotherapy

Allergen immunotherapy (AIT) is also known as desensitization and can help to switch off an allergy over time. It's an effective treatment for allergies to the stings of bees, wasps and fire ants. Your child will be given insect venom in gradually increasing doses. This allows their immune system to become tolerant, so they no longer have allergic responses.

AIT is not helpful for children who have large local swellings alone, and it may not be necessary in children with isolated rashes. A history of anaphylaxis to an insect sting, however, would be an appropriate reason to begin AIT. Your child would need to be evaluated by an allergist before AIT is considered, and many allergists often wait until the child is at least five years old before starting AIT, but it depends on the individual child and their situation. The duration of treatment is usually three to five years.

WHAT YOU CAN DO AS A PARENT

- If your child has a severe allergy, always carry an Anaphylaxis Emergency Plan and an adrenaline auto-injector (EpiPen) that is readily available to treat anaphylaxis if your child is bitten or stung. Food Allergy Canada has a very good Anaphylaxis Emergency Plan as a download on its website (foodallergycanada.ca).
- Place a medical alert bracelet on your child, which will increase the likelihood that adrenaline is administered in an emergency.

- Avoid giving your child medication that may increase the severity of anaphylaxis or complicate its treatment. Beta-blockers and ACE inhibitors (heart medication that helps relax your veins and arteries to lower your blood pressure) fall into this group.
- Seek urgent medical assistance for your child if they are stung or bitten.
- Teach your child not to disturb insect nests or irritate insects. Instruct your child to seek shelter if they encounter the insect that they are allergic to outside. Honeybees normally only sting in self-defense.
- Dress your child in light-colored clothing that covers much of their body, and make sure that they wear shoes. Stings often occur on bare feet, so children with allergies to bites or stings should try to always wear shoes when outdoors.
- Wasps are generally more aggressive than bees, and they are attracted to food and drink. Teach your child to always check their drink when outdoors.
- Where possible, drive with the windows up.
- If you find an insect nest on your property, have it removed by professionals.

PENICILLIN ALLERGY

A penicillin allergy is the most common drug allergy and a frequent cause of skin rashes. The good news is that most children with an immediate or mild non-immediate reaction to penicillin antibiotics outgrow it with time. Up to 80 percent of kids with penicillin allergy will have outgrown it by the time they are ten years old.

Diagnosing penicillin allergy

Unfortunately, penicillin allergy can be harder to diagnose than most parents expect. A study published in the journal *Pediatrics* found that less than 1 percent of children whose parents said they had penicillin allergy based on family history—or what the researchers referred to as low-risk symptoms (rash, itching, vomiting, diarrhea, runny nose and cough)—were actually allergic to penicillin.

It can be challenging to determine the cause of rashes in young children, as they are very common. For example, viral rashes can often be confused with antibiotic reactions. Additionally, some drug reactions can be delayed, and thus a skin rash may not be attributed to the effects of an antibiotic.

Blood tests can give clues to but not confirm an allergic reaction to a drug. For immediate reactions, validated skin testing only exists for penicillin and not for any of the other antibiotics.

Symptoms

Penicillin allergies can cause life-threatening reaction; fortunately, most children with penicillin allergy have milder reactions, leading to simple skin rashes such as hives.

Children with more severe allergic symptoms will experience hives as well as wheezing, difficulty breathing or swallowing, and/or swelling in their mouth or throat. Anaphylaxis can also occur.

Management

In the case of confirmed penicillin allergy, strict avoidance of penicillin is recommended. If your child is allergic to penicillin, then they should also avoid Amoxil, Augmentin and any other penicillin-like antibiotics.

If the future use of penicillin is necessary to treat a specific infection, then desensitization may be considered by an allergist. Desensitization is a specific form of allergen immunotherapy (AIT) where the child will be given penicillin in increasing doses over a short period of time to allow their immune system to become tolerant so that they can use the medication safely. The tolerance is temporary; once the treatment stops, the allergic reaction returns. In general, desensitization is only used in a situation where there is no alternative medication available. It would not to be used for cases where there have been severe skin reactions to penicillin.

"Delabeling" penicillin allergies in children

The label of "penicillin allergy" can be associated with adverse health outcomes for children. If the medical practitioners who are treating your child mistakenly believe that your child has penicillin allergy, your child will be prescribed alternative broad-spectrum antibiotics—and this can lead to an increase in the risk of antibiotic resistance. Wherever possible, penicillin allergy should be confirmed by an allergist. If your child is found not to have penicillin allergy, let your doctors and health-care professionals know so that they can update their records and treat your child accordingly.

ASPIRIN AND NSAID ALLERGY

Aspirin is a well-known and long-used drug that can reduce pain from inflammation, injury and fever. A number of similar synthetic non-steroidal anti-inflammatory drugs (NSAIDs) have been developed, such as Voltaren, Nurofen, Naprosyn, Celebrex, Indocid, Feldene and Mobic. Aspirin or other NSAIDs may be present in over-the-counter painkillers and found in medications for headache, period pain and sinus pain. They are also found in cold and flu tablets.

Children should not take aspirin, as it has been linked to Reye syndrome—an extremely rare but serious illness that can affect the brain and liver. Reye syndrome is most common in kids who are recovering from a viral infection.

Testing for NSAID sensitivity

There is no reliable skin-prick test or blood test to confirm or exclude sensitivity to aspirin and NSAIDs. The only way to check sensitivity is to ingest the drug, starting with a very low dose, under strict medical supervision and monitoring for a reaction. Challenge testing is not always necessary, but may be advised in some circumstances to prove that sensitivity exists.

Symptoms

Symptoms of aspirin and NSAID allergies include flushing, hives, blocked and runny nose and wheezing, usually within an hour of taking a tablet. If a child has hives, nasal polyps or asthma, their risk of an aspirin allergy is 10–30 percent, compared to

1 percent in children without these conditions. These reactions can also be triggered by non–aspirin NSAIDs.

Management

For children who react to aspirin or NSAIDs, the treatment is complete avoidance. It is very important for parents to read all labels of both prescription and over-the-counter medications, since many contain variations of either aspirin or NSAIDs. If you have any doubts, check with your doctor or pharmacist.

PET ALLERGY

Pet allergies occur after a child develops IgE (allergy) antibodies to allergens from an animal. There is evidence that having a pet can reduce the risk of future allergies in your children (see Chapter 11), but this section talks about what happens when there's *already* confirmed pet allergy.

Which pets usually trigger an allergy?

The most common allergies are to cats and dogs. All breeds of cat and dog can potentially cause allergies, although some may shed less hair and skin particles (and therefore allergens) than others.

Cats

The main source of cat allergen is the sebaceous glands in the cat's skin. Cats often lick themselves, and this helps to spread cat allergen, which is sticky and glues itself to dander (particles shed from skin, hair and fur), dust particles and all parts of the home.

As all cats have sebaceous glands, all cat breeds can potentially cause allergies.

Cat allergen can remain distributed throughout the home for up to six months and in the cat's bedding for up to four years. The allergen spreads so thoroughly that it can be measured in the homes of non-pet owners and on the clothing of co-workers of cat owners. Cat allergen has even been detected in the Antarctic, even though cats have never been there.

Dogs

The main sources of dog allergen are their dander and saliva. Dander in particular can spread the allergen. All dog breeds cause allergies, although some do not shed as much dander (and therefore allergen).

Other animals

Although not as common as cat and dog allergies, some children may be allergic to other animals including horses, mice, rats, rabbits, guinea pigs and birds.

Symptoms

Common symptoms of pet allergies are itchy eyes, sneezing, itchy nose and sometimes wheezing or asthma attacks. Hives can occur where the pet has licked the child, but anaphylaxis to pet allergens is very rare.

For many pet-allergic children, symptoms occur quickly after exposure to the pet. If your child has symptoms suggestive of allergy to pets, your allergist can confirm the presence of allergy antibodies using a skin-prick test or blood test.

WHAT YOU CAN DO AS A PARENT

- If your child has pet allergy, do not bring a furred animal into the home.
- If symptoms in your child are severe, find your existing pet a new home and remove its bedding from your house.
- Do not smoke, as exposure to environmental smoke worsens allergic symptoms. Smoking also makes a number of allergies, including pet allergies, more likely to develop.

CHAPTER 9
NINE-WEEK INFANT MEAL PLAN

In previous chapters, we looked at the evidence for preventing allergies—and food allergies in particular. As a doctor and a parent, I think it's important to have an optimistic outlook towards preventing allergies in your baby. You should feel empowered by the idea that you can help to prevent allergies before they arise, and make the most of the narrow window in which you can do so.

We've already covered the groundbreaking LEAP study, which proved that introducing peanut to babies at high risk of peanut allergy reduced the odds of that allergy appearing by 70 percent. A year later, in 2016, the same UK researchers published the EAT study, which looked at whether or not it is safe to introduce six allergenic foods (peanut, egg, milk, sesame, whitefish and wheat) to infants. Turns out, it was. There was zero anaphylaxis among the 1303 participating babies, and the rate of allergy was reduced in those babies whose parents were able to feed them the six foods every week. Unfortunately, only a handful of patients were able to stick to the regimen of feeding all six foods to their baby every week.

ABOUT THE MEAL PLAN

In this chapter you'll find a meal plan that will introduce the nine major food allergens in nine weeks. The meal plan should be adopted during the critical window of between six and twelve months of age. Ideally, prior to the introduction of solids at six months, babies should be breastfed exclusively.

This meal plan introduces each food allergen one at a time. The plan proposes that you introduce a new food allergen to your baby on the Monday of each week (highlighted in bold on the plan), and then you repeat this allergen introduction over the following three days to make sure that there is no reaction. That allergen then becomes incorporated into the standard meal plan for your baby and is consumed on a regular basis—three times a week.

Non-allergenic foods can be offered at any time, and soon you will see an increase in the diversity of your baby's diet. In the meal plan, I've included some non-allergenic foods—such as pureed fruit or pureed steamed carrot—which you can prepare for your baby. Alternatively, you can use any packaged pureed food from the supermarket as a substitute. For vegetarians, any meat in the meal plan can be replaced with tofu or legumes as a source of protein for your baby.

Some sensible precautions need to be taken when using this meal plan, such as not giving whole nuts to infants because of the choking hazard. As mentioned in Chapter 5, babies need to be developmentally ready to eat solid foods. They must be able to hold their head and neck up straight, reach out for food and open their mouth when offered food on a spoon. Be careful not to overfeed your baby. They will know when enough is enough, and will turn away and lose interest in the food if they're full.

In Chapter 7, we talked about how to introduce peanuts to babies, starting with a quarter of a teaspoon. You should use the same technique for introducing each of the food allergens.

Important advice

If your baby develops a reaction to a particular food, do not give them any more of that food. This food is now a suspected food allergen for your baby, and you should seek the advice of an allergist, who may wish to do some tests. In general, you should wait at least three months before introducing the food allergen to your baby again. At that stage, your baby may be able to handle the food allergen in a different texture (for example, mashed or grated rather than pureed).

When following the meal plan, avoid starting a new allergenic food if your baby is on antibiotics. It's okay to delay until the course of antibiotics is over—we have a whole six-month window to introduce these foods.

I would like to acknowledge the generous contribution of Geraldine Georgeou, an accredited practicing dietitian from Designer Diets, who reviewed this infant meal plan. The advice in this chapter should be considered alongside the other lifestyle and environmental information provided in this book.

WEEK 1—INTRODUCING EGG

	Early feed	Breakfast/ morning snack	Lunchtime	Afternoon snack	Dinner	Bedtime
MONDAY	Breastmilk or infant formula feed	Iron-fortified infant wholegrain cereal with pureed banana, formula, breastmilk or cooled boiled tap water can be mixed with the cereal	Pureed fruit, such as banana, kiwi, cooked apple and/or cooked pear **Introduce a small amount of mashed egg** Drinks: breastmilk, infant formula, cooled boiled tap water (as required)	Pureed steamed carrot Drinks: breastmilk, infant formula, cooled boiled tap water (as required)	Pureed cooked chicken; pureed cooked pumpkin, carrot and zucchini (not mixed together*); pureed cooked rice	Breastmilk or infant formula feed
TUESDAY	Breastmilk or infant formula feed	Iron-fortified infant wholegrain cereal with pureed sweet potato; formula, breastmilk or cooled boiled tap water can be mixed with the cereal	Pureed fruit, such as banana, kiwi, cooked apple and/or cooked pear **Introduce a small amount of mashed egg** Drinks: breastmilk, infant formula, cooled boiled tap water (as required)	Pureed steamed carrot Drinks: breastmilk, infant formula, cooled boiled tap water (as required)	Pureed cooked beef; pureed cooked carrot, cauliflower and spinach (not mixed together)	Breastmilk or infant formula feed

* I suggest keeping cooked veggies separate so the baby can distinguish each food and its individual flavor.

WEEK 1—INTRODUCING EGG

	Early feed	Breakfast/morning snack	Lunchtime	Afternoon snack	Dinner	Bedtime
WEDNESDAY	Breastmilk or infant formula feed	Iron-fortified infant wholegrain cereal with pureed pear; formula, breastmilk or cooled boiled tap water can be mixed with the cereal	Pureed fruit, such as banana, kiwi, cooked apple and/or cooked pear **Introduce a small amount of mashed egg** Drinks: breastmilk, infant formula, cooled boiled tap water (as required)	Pureed steamed carrot Drinks: breastmilk, infant formula, cooled boiled tap water (as required)	Pureed beef stew with pureed steamed carrot	Breastmilk or infant formula feed .
THURSDAY	Breastmilk or infant formula feed	Iron-fortified infant wholegrain cereal with pureed apple; formula, breastmilk or cooled boiled tap water can be mixed with the cereal	Pureed fruit, such as banana, kiwi, cooked apple and/or cooked pear **Introduce a small amount of mashed egg** Drinks: breastmilk, infant formula, cooled boiled tap water (as required)	Pureed steamed carrot Drinks: breastmilk, infant formula, cooled boiled tap water (as required)	Pureed cooked chicken with pureed sweet corn and steamed carrot	Breastmilk or infant formula feed

WEEK 1–INTRODUCING EGG

	Early feed	Breakfast/ morning snack	Lunchtime	Afternoon snack	Dinner	Bedtime
FRIDAY	Breastmilk or infant formula feed	Iron-fortified infant wholegrain cereal with creamed vegetables or pureed banana; formula, breastmilk or cooled boiled tap water can be mixed with the cereal	Pureed fruit, such as banana, kiwi, cooked apple and/or cooked pear Drinks: breastmilk, infant formula, cooled boiled tap water (as required)	Pureed steamed carrot Drinks: breastmilk, infant formula, cooled boiled tap water (as required)	Pureed roast chicken and vegetable stew	Breastmilk or infant formula feed
SATURDAY	Breastmilk or infant formula feed	Iron-fortified infant wholegrain cereal with pureed mango; formula, breastmilk or cooled boiled tap water can be mixed with the cereal	Pureed fruit, such as banana, kiwi, cooked apple and/or cooked pear Drinks: breastmilk, infant formula, cooled boiled tap water (as required)	Pureed steamed carrot; **a small amount of mashed egg** Drinks: breastmilk, infant formula, cooled boiled tap water (as required)	Root vegetable soup with pureed cooked chicken	Breastmilk or infant formula feed

WEEK 1–INTRODUCING EGG

	Early feed	Breakfast/ morning snack	Lunchtime	Afternoon snack	Dinner	Bedtime
SUNDAY	Breastmilk or infant formula feed	Iron-fortified infant wholegrain cereal with pureed sweet potato and/ or apple; formula, breast-milk or cooled boiled tap water can be mixed with the cereal	Pureed fruit, such as banana, kiwi, cooked apple and/or cooked pear Drinks: breastmilk, infant formula, cooled boiled tap water (as required)	Pureed steamed carrot Drinks: breastmilk, infant formula, cooled boiled tap water (as required)	Baby ratatouille with pureed cooked lamb	Breastmilk or infant formula feed

WEEK 2—INTRODUCING DAIRY

	Early feed	Breakfast/ morning snack	Lunchtime	Afternoon snack	Dinner	Bedtime
MONDAY	Breastmilk or infant formula feed	Iron-fortified infant wholegrain cereal with pureed banana; formula, breastmilk or cooled boiled tap water can be mixed with the cereal	Pureed fruit, such as banana, kiwi, cooked apple and/or cooked pear **Introduce a small amount of full-fat yoghurt** Drinks: breastmilk, infant formula, cooled boiled tap water (as required)	Pureed steamed carrot; mashed egg Drinks: breastmilk, infant formula, cooled boiled tap water (as required)	Pureed cooked chicken; pureed cooked pumpkin, carrot and zucchini (not mixed together); pureed cooked rice	Breastmilk or infant formula feed
TUESDAY	Breastmilk or infant formula feed	Iron-fortified infant wholegrain cereal with pureed sweet potato; formula, breastmilk or cooled boiled tap water can be mixed with the cereal	Pureed fruit, such as banana, kiwi, cooked apple and/or cooked pear **Introduce a small amount of full-fat yoghurt** Drinks: breastmilk, infant formula, cooled boiled tap water (as required)	Pureed steamed carrot Drinks: breastmilk, infant formula, cooled boiled tap water (as required)	Pureed cooked carrot, cauliflower and spinach (not mixed together); a small amount of mashed egg	Breastmilk or infant formula feed

WEEK 2—INTRODUCING DAIRY

	Early feed	Breakfast/ morning snack	Lunchtime	Afternoon snack	Dinner	Bedtime
WEDNESDAY	Breastmilk or infant formula feed	Iron-fortified infant wholegrain wholegrain cereal with pureed pear; formula, breastmilk or cooled boiled tap water can be mixed with the cereal	Pureed fruit, such as banana, kiwi, cooked apple and/or cooked pear **Introduce a small amount of full-fat yoghurt** Drinks: breastmilk, infant formula, cooled boiled tap water (as required)	Pureed steamed carrot; mashed egg Drinks: breastmilk, infant formula, cooled boiled tap water (as required)	Pureed beef stew with pureed steamed carrot	Breastmilk or infant formula feed
THURSDAY	Breastmilk or infant formula feed	Iron-fortified infant wholegrain cereal with pureed pear; formula, breastmilk or cooled boiled tap water can be mixed with the cereal	Pureed fruit, such as banana, kiwi, cooked apple and/or cooked pear **Introduce a small amount of full-fat yoghurt** Drinks: breastmilk, infant formula, cooled boiled tap water (as required)	Pureed steamed carrot Drinks: breastmilk, infant formula, cooled boiled tap water (as required)	Pureed cooked chicken with pureed sweet corn and pear; a small amount of mashed egg	Breastmilk or infant formula feed

WEEK 2—INTRODUCING DAIRY

	Early feed	Breakfast/ morning snack	Lunchtime	Afternoon snack	Dinner	Bedtime
FRIDAY	Breastmilk or infant formula feed	Iron-fortified infant wholegrain cereal with creamed vegetables (such as pumpkin); formula, breastmilk or cooled boiled tap water can be mixed with the cereal	Pureed fruit, such as banana, kiwi, cooked apple and/or cooked pear. Drinks: breastmilk, infant formula, cooled boiled tap water (as required)	Pureed steamed carrot; mashed egg. Drinks: breast milk, infant formula, cooled boiled tap water (as required)	Root vegetable soup with pureed cooked chicken	Breastmilk or infant formula feed
SATURDAY	Breastmilk or infant formula feed	Iron-fortified infant wholegrain cereal with pureed mango; formula, breastmilk or cooled boiled tap water can be mixed with the cereal	Pureed fruit, such as banana, kiw, cooked apple and/or cooked pear; a small amount of **full-fat yoghurt**. Drinks: breastmilk, infant formula, cooled boiled tap water (as required)	Pureed steamed carrot. Drinks: breastmilk, infant formula, cooled boiled tap water (as required)	Pureed roast chicken and vegetable stew; a small amount of mashed egg	Breastmilk or infant formula feed

WEEK 2—INTRODUCING DAIRY

	Early feed	Breakfast/ morning snack	Lunchtime	Afternoon snack	Dinner	Bedtime
SUNDAY	Breastmilk or infant formula feed	Iron-fortified infant wholegrain cereal with pureed sweet potato and/or apple; formula, breastmilk or cooled boiled tap water can be mixed with the cereal	Pureed fruit, such as banana, kiwi, cooked apple and/or cooked pear Drinks: breastmilk, infant formula, cooled boiled tap water (as required)	Pureed steamed carrot Drinks: breast milk, infant formula, cooled boiled tap water (as required)	Baby ratatouille with pureed cooked lamb	Breastmilk or infant formula feed

WEEK 3—INTRODUCING SESAME

	Early feed	Breakfast/ morning snack	Lunchtime	Afternoon snack	Dinner	Bedtime
MONDAY	Breastmilk or infant formula feed	Iron-fortified infant wholegrain cereal with pureed banana; formula, breastmilk or cooled boiled tap water can be mixed with the cereal	Pureed fruit, such as banana, kiwi, cooked apple and/or cooked pear; full-fat yoghurt **Introduce a small amount of hummus dip** Drinks: breastmilk, infant formula, cooled boiled tap water (as required)	Pureed steamed carrot Drinks: breastmilk, infant formula, cooled boiled tap water (as required)	Pureed cooked chicken; pureed cooked pumpkin, carrot and zucchini (not mixed together); pureed cooked rice	Breastmilk or infant formula feed
TUESDAY	Breastmilk or infant formula feed	Iron-fortified infant wholegrain cereal with pureed sweet potato; formula, breastmilk or cooled boiled tap water can be mixed with the cereal	Pureed fruit, such as banana, kiwi, cooked apple and/or cooked pear; full-fat yoghurt **Introduce a small amount of hummus dip** Drinks: breastmilk, infant formula, cooled boiled tap water (as required)	Pureed steamed carrot Drinks: breastmilk, infant formula, cooled boiled tap water (as required)	Pureed cooked beef; pureed cooked carrot, cauliflower and spinach (not mixed together)	Breastmilk or infant formula feed

WEEK 3—INTRODUCING SESAME

	Early feed	Breakfast/ morning snack	Lunchtime	Afternoon snack	Dinner	Bedtime
WEDNESDAY	Breastmilk or infant formula feed	Iron-fortified infant wholegrain cereal with pureed banana; formula, breastmilk or cooled boiled tap water can be mixed with the cereal	Pureed fruit, such as banana, kiwi, cooked apple and/or cooked pear; full-fat yoghurt **Introduce a small amount of hummus dip** Drinks: breastmilk, infant formula, cooled boiled tap water (as required)	Pureed steamed carrot Drinks: breastmilk, infant formula, cooled boiled tap water (as required)	Pureed beef stew with pureed steamed carrot	Breastmilk or infant formula feed
THURSDAY	Breastmilk or infant formula feed	Iron-fortified infant wholegrain cereal with pureed pear; formula, breastmilk or cooled boiled tap water can be mixed with the cereal	Pureed fruit, such as banana, kiwi, cooked apple and/or cooked pear; full-fat yoghurt **Introduce a small amount of hummus dip** Drinks: breastmilk, infant formula, cooled boiled tap water (as required)	Pureed steamed carrot Drinks: breastmilk, infant formula, cooled boiled tap water (as required)	Pureed cooked chicken with pureed sweet corn and potato	Breastmilk or infant formula feed

WEEK 3–INTRODUCING SESAME

	Early feed	Breakfast/ morning snack	Lunchtime	Afternoon snack	Dinner	Bedtime
FRIDAY	Breastmilk or infant formula feed	Iron-fortified infant wholegrain cereal with creamed vegetables; formula, breastmilk or cooled boiled tap water can be mixed with the cereal	Pureed fruit such as banana, kiwi, cooked apple and/or cooked pear; full-fat yoghurt Drinks: breastmilk, infant formula, cooled boiled tap water (as required)	Pureed steamed carrot Drinks: breastmilk, infant formula, cooled boiled tap water (as required)	Pureed roast chicken and vegetable stew	Breastmilk or infant formula feed
SATURDAY	Breastmilk or infant formula feed	Iron-fortified infant wholegrain cereal with pureed mango; formula, breastmilk or cooled boiled tap water can be mixed with the cereal	Pureed fruit, such as banana, kiwi, cooked apple and/or cooked pear; full-fat yoghurt Drinks: breastmilk, infant formula, cooled boiled tap water (as required)	Pureed steamed carrot; a small amount of hummus dip Drinks: breastmilk, infant formula, cooled boiled tap water (as required)	Root vegetable soup with pureed cooked chicken	Breastmilk or infant formula feed

WEEK 3—INTRODUCING SESAME

	Early feed	Breakfast/ morning snack	Lunchtime	Afternoon snack	Dinner	Bedtime
SUNDAY	Breastmilk or infant formula feed	Iron-fortified infant wholegrain cereal with pureed sweet potato and/ or apple; formula, breast-milk or cooled boiled tap water can be mixed with the cereal	Pureed fruit, such as banana, kiwi, cooked apple and/or cooked pear; full-fat yoghurt Drinks: breastmilk, infant formula, cooled boiled tap water (as required)	Pureed steamed carrot Drinks: breastmilk, infant formula, cooled boiled tap water (as required)	Baby ratatouille with pureed cooked lamb	Breastmilk or infant formula feed

WEEK 4–INTRODUCING FISH

	Early feed	Breakfast/ morning snack	Lunchtime	Afternoon snack	Dinner	Bedtime
MONDAY	Breastmilk or infant formula feed	Iron-fortified infant wholegrain cereal with pureed banana; formula, breastmilk or cooled boiled tap water can be mixed with the cereal	Pureed fruit, such as banana, kiwi, cooked apple and/or cooked pear; full-fat yoghurt **Introduce a small amount of steamed boneless fish** Drinks: breastmilk, infant formula, cooled boiled tap water (as required)	Pureed steamed carrot; hummus dip Drinks: breastmilk, infant formula, cooled boiled tap water (as required)	Pureed cooked chicken; pureed cooked pumpkin, carrot and zucchini (not mixed together); pureed cooked rice	Breastmilk or infant formula feed
TUESDAY	Breastmilk or infant formula feed	Iron-fortified infant wholegrain cereal with pureed sweet potato or pumpkin; formula, breastmilk or cooled boiled tap water can be mixed with the cereal	Pureed fruit, such as banana, kiwi, cooked apple and/or cooked pear; full-fat yoghurt **Introduce a small amount of steamed boneless fish** Drinks: breastmilk, infant formula, cooled boiled tap water (as required)	Pureed steamed carrot; mashed egg Drinks: breastmilk, infant formula, cooled boiled tap water (as required)	Pureed beef; pureed cooked carrot, cauliflower and spinach (not mixed together)	Breastmilk or infant formula feed

WEEK 4—INTRODUCING FISH

	Early feed	Breakfast/ morning snack	Lunchtime	Afternoon snack	Dinner	Bedtime
WEDNESDAY	Breastmilk or infant formula feed	Iron-fortified infant wholegrain cereal with pureed pear; formula, breastmilk or cooled boiled tap water can be mixed with the cereal	Pureed fruit, such as banana, kiwi, cooked apple and/or cooked pear; full-fat yoghurt **Introduce a small amount of steamed boneless fish** Drinks: breastmilk, infant formula, cooled boiled tap water (as required)	Pureed steamed carrot; hummus dip Drinks: breastmilk, infant formula, cooled boiled tap water (as required)	Pureed beef stew with pureed steamed carrot	Breastmilk or infant formula feed
THURSDAY	Breastmilk or infant formula feed	Iron-fortified infant wholegrain cereal with pureed pear; formula, breastmilk or cooled boiled tap water can be mixed with the cereal	Pureed fruit, such as banana, kiwi, cooked apple and/or cooked pear; full-fat yoghurt **Introduce a small amount of steamed boneless fish** Drinks: breastmilk, infant formula, cooled boiled tap water (as required)	Pureed steamed carrot; mashed egg Drinks: breastmilk, infant formula, cooled boiled tap water (as required)	Pureed cooked chicken with pureed sweet corn and potato	Breastmilk or infant formula feed

WEEK 4—INTRODUCING FISH

	Early feed	Breakfast/ morning snack	Lunchtime	Afternoon snack	Dinner	Bedtime
FRIDAY	Breastmilk or infant formula feed	Iron-fortified infant wholegrain cereal with creamed vegetables; formula, breastmilk or cooled boiled tap water can be mixed with the cereal	Pureed fruit, such as banana, kiwi, cooked apple and/or cooked pear; full-fat yoghurt Drinks: breastmilk, infant formula, cooled boiled tap water (as required)	Pureed steamed carrot; hummus dip Drinks: breastmilk, infant formula, cooled boiled tap water (as required)	Pureed roast chicken and vegetable stew	Breastmilk or infant formula feed
SATURDAY	Breastmilk or infant formula feed	Iron-fortified infant wholegrain cereal with pureed mango; formula, breastmilk or cooled boiled tap water can be mixed with the cereal	Pureed fruit, such as banana, kiwi, cooked apple and/or cooked pear; full-fat yoghurt Drinks: breastmilk, infant formula, cooled boiled tap water (as required)	Pureed steamed carrot; mashed egg Drinks: breastmilk, infant formula, cooled boiled tap water (as required)	Root vegetable soup with a small amount of steamed boneless fish	Breastmilk or infant formula feed

WEEK 4—INTRODUCING FISH

	Early feed	Breakfast/ morning snack	Lunchtime	Afternoon snack	Dinner	Bedtime
SUNDAY	Breastmilk or infant formula feed	Iron-fortified infant wholegrain cereal with pureed apple; formula, breastmilk or cooled boiled tap water can be mixed with the cereal	Pureed fruit, such as banana, kiwi, cooked apple and/or cooked pear, full-fat yoghurt Drinks: breastmilk, infant formula, cooled boiled tap water (as required)	Pureed steamed carrot Drinks: breastmilk, infant formula, cooled boiled tap water (as required)	Baby ratatouille with pureed cooked lamb	Breastmilk or infant formula feed

WEEK 5—INTRODUCING PEANUT

	Early feed	Breakfast/ morning snack	Lunchtime	Afternoon snack	Dinner	Bedtime
MONDAY	Breastmilk or infant formula feed	Iron-fortified infant wholegrain cereal with pureed banana; formula, breastmilk or cooled boiled tap water can be mixed with the cereal	Pureed fruit, such as banana, kiwi, cooked apple and/or cooked pear; full-fat yoghurt **Introduce a small amount of smooth peanut butter** Drinks: breastmilk, infant formula, cooled boiled tap water (as required)	Pureed steamed carrot; hummus dip Drinks: breastmilk, infant formula, cooled boiled tap water (as required)	Pureed cooked chicken; pureed cooked pumpkin, carrot and zucchini (not mixed together); pureed cooked rice	Breastmilk or infant formula feed
TUESDAY	Breastmilk or infant formula feed	Iron-fortified infant wholegrain cereal with pureed sweet potato or pumpkin; formula, breastmilk or cooled boiled tap water can be mixed with the cereal	Pureed fruit, such as banana, kiwi, cooked apple and/or cooked pear; full-fat yoghurt **Introduce a small amount of smooth peanut butter** Drinks: breastmilk, infant formula, cooled boiled tap water (as required)	Pureed steamed carrot; mashed egg Drinks: breastmilk, infant formula, cooled boiled tap water (as required)	Pureed cooked carrot, cauliflower and spinach (not mixed together); steamed boneless fish	Breastmilk or infant formula feed

WEEK 5—INTRODUCING PEANUT

	Early feed	Breakfast/morning snack	Lunchtime	Afternoon snack	Dinner	Bedtime
WEDNESDAY	Breastmilk or infant formula feed	Iron-fortified infant wholegrain cereal with pureed pear; formula, breastmilk or cooled boiled tap water can be mixed with the cereal	Pureed fruit, such as banana, kiwi, cooked apple and/or cooked pear; full-fat yoghurt **Introduce a small amount of smooth peanut butter** Drinks: breastmilk, infant formula, cooled boiled tap water (as required)	Pureed steamed carrot; hummus dip Drinks: breastmilk, infant formula, cooled boiled tap water (as required)	Pureed beef stew with pureed steamed carrot	Breastmilk or infant formula feed
THURSDAY	Breastmilk or infant formula feed	Iron-fortified infant wholegrain cereal with pureed pear; formula, breastmilk or cooled boiled tap water can be mixed with the cereal	Pureed fruit, such as banana, kiwi, cooked apple and/or cooked pear; full-fat yoghurt **Introduce a small amount of smooth peanut butter** Drinks: breastmilk, infant formula, cooled boiled tap water (as required)	Pureed steamed carrot; mashed egg Drinks: breastmilk, infant formula, cooled boiled tap water (as required)	Steamed boneless fish with pureed sweet corn and mashed potato	Breastmilk or infant formula feed

WEEK 5—INTRODUCING PEANUT

	Early feed	Breakfast/ morning snack	Lunchtime	Afternoon snack	Dinner	Bedtime
FRIDAY	Breastmilk or infant formula feed	Iron-fortified infant wholegrain cereal with creamed vegetables; formula, breastmilk or cooled boiled tap water can be mixed with the cereal	Pureed fruit, such as banana, kiwi, cooked apple and/or cooked pear; full-fat yoghurt Drinks: breastmilk, infant formula, cooled boiled tap water (as required)	Pureed steamed carrot; hummus dip Drinks: breastmilk, infant formula, cooled boiled tap water (as required)	Pureed roast chicken and vegetable stew	Breastmilk or infant formula feed
SATURDAY	Breastmilk or infant formula feed	Iron-fortified infant wholegrain cereal with pureed mango; formula, breastmilk or cooled boiled tap water can be mixed with the cereal **Introduce a small amount of smooth peanut butter**	Pureed fruit, such as banana, kiwi, cooked apple and/or cooked pear; full-fat yoghurt Drinks: breastmilk, infant formula, cooled boiled tap water (as required)	Pureed steamed carrot; mashed egg Drinks: breastmilk, infant formula, cooled boiled tap water (as required)	Root vegetable soup; steamed boneless fish	Breastmilk or infant formula feed

WEEK 5—INTRODUCING PEANUT

	Early feed	Breakfast/ morning snack	Lunchtime	Afternoon snack	Dinner	Bedtime
SUNDAY	Breastmilk or infant formula feed	Iron-fortified infant wholegrain cereal with pureed sweet potato and/or apple; formula, breastmilk or cooled boiled tap water can be mixed with the cereal	Pureed fruit, such as banana, kiwi, cooked apple and/or cooked pear; full-fat yoghurt Drinks: breastmilk, infant formula, cooled boiled tap water (as required)	Pureed steamed carrot Drinks: breastmilk, infant formula, cooled boiled tap water (as required)	Baby ratatouille with pureed cooked lamb	Breastmilk or infant formula feed

WEEK 6—INTRODUCING WHEAT

	Early feed	Breakfast/ morning snack	Lunchtime	Afternoon snack	Dinner	Bedtime
MONDAY	Breastmilk or infant formula feed	Iron-fortified infant wholegrain cereal with pureed banana; formula, breastmilk or cooled boiled tap water can be mixed with the cereal Smooth peanut butter	Pureed fruit, such as banana, kiwi, cooked apple and/or cooked pear; full-fat yoghurt **Introduce a small amount of wheat-based porridge or soft Weetabix** Drinks: breastmilk, infant formula, cooled boiled tap water (as required)	Pureed steamed carrot; hummus dip Drinks: breastmilk, infant formula, cooled boiled tap water (as required)	Pureed steamed carrot and mashed potato; steamed boneless fish	Breastmilk or infant formula feed
TUESDAY	Breastmilk or infant formula feed	Iron-fortified infant wholegrain cereal with pureed sweet potato or pumpkin; formula, breastmilk or cooled boiled tap water can be mixed with the cereal	Pureed fruit, such as banana, kiwi, cooked apple and/or cooked pear; full-fat yoghurt **Introduce a small amount of wheat-based porridge or soft Weetabix** Drinks: breastmilk, infant formula, cooled boiled tap water (as required)	Pureed steamed carrot; mashed egg Drinks: breastmilk, infant formula, cooled boiled tap water (as required)	Pureed cooked chicken; pureed cooked carrot, cauliflower and spinach (not mixed together)	Breastmilk or infant formula feed

WEEK 6—INTRODUCING WHEAT

	Early feed	Breakfast/ morning snack	Lunchtime	Afternoon snack	Dinner	Bedtime
WEDNESDAY	Breastmilk or infant formula feed	Iron-fortified infant wholegrain cereal with pureed banana; formula, breastmilk or cooled boiled tap water can be mixed with the cereal Smooth peanut butter	Pureed fruit, such as banana, kiwi, cooked apple and/or cooked pear; full-fat yoghurt **Introduce a small amount of wheat-based porridge or soft Weetabix** Drinks: breastmilk, infant formula, cooled boiled tap water (as required)	Pureed steamed carrot; hummus dip Drinks: breastmilk, infant formula, cooled boiled tap water (as required)	Pureed steamed carrot and mashed potato; steamed bone-less fish	Breastmilk or infant formula feed
THURSDAY	Breastmilk or infant formula feed	Iron-fortified infant wholegrain cereal with pureed pear; formula, breastmilk or cooled boiled tap water can be mixed with the cereal	Pureed fruit, such as banana, kiwi, cooked apple and/or cooked pear; full-fat yoghurt **Introduce a small amount of wheat-based porridge or soft Weetabix** Drinks: breastmilk, infant formula, cooled boiled tap water (as required)	Pureed steamed carrot; mashed egg Drinks: breastmilk, infant formula, cooled boiled tap water (as required)	Pureed cooked chicken with pureed sweet corn and potato	Breastmilk or infant formula feed

WEEK 6—INTRODUCING WHEAT

	Early feed	Breakfast/ morning snack	Lunchtime	Afternoon snack	Dinner	Bedtime
FRIDAY	Breastmilk or infant formula feed	Iron-fortified infant wholegrain cereal with creamed vegetables; formula, breastmilk or cooled boiled tap water can be mixed with the cereal Smooth peanut butter	Pureed fruit, such as banana, kiwi, cooked apple and/or cooked pear; full-fat yoghurt Drinks: breastmilk, infant formula, cooled boiled tap water (as required)	Pureed steamed carrot; hummus dip Drinks: breastmilk, infant formula, cooled boiled tap water (as required)	Steamed boneless fish and vegetable stew	Breastmilk or infant formula feed
SATURDAY	Breastmilk or infant formula feed	Iron-fortified infant wholegrain cereal with pureed mango; formula, breastmilk or cooled boiled tap water can be mixed with the cereal Introduce a small amount of wheat-based porridge or soft Weetabix	Pureed fruit, such as banana, kiwi, cooked apple and/or cooked pear; full-fat yoghurt Drinks: breastmilk, infant formula, cooled boiled tap water (as required)	Pureed steamed carrot; mashed egg Drinks: breastmilk, infant formula, cooled boiled tap water (as required)	Root vegetable soup and pureed cooked lamb	Breastmilk or infant formula feed

WEEK 6—INTRODUCING WHEAT

	Early feed	Breakfast/ morning snack	Lunchtime	Afternoon snack	Dinner	Bedtime
SUNDAY	Breastmilk or infant formula feed	Iron-fortified infant wholegrain cereal with pureed sweet potato and/ or apple; formula, breast-milk or cooled boiled tap water can be mixed with the cereal	Pureed fruit, such as banana, kiwi, cooked apple and/or cooked pear, full-fat yoghurt Drinks: breastmilk, infant formula, cooled boiled tap water (as required)	Pureed steamed carrot Drinks: breastmilk, infant formula, cooled boiled tap water (as required)	Baby ratatouille with pureed cooked lamb	Breastmilk or infant formula feed

WEEK 7–INTRODUCING SHELLFISH

	Early feed	Breakfast/ morning snack	Lunchtime	Afternoon snack	Dinner	Bedtime
MONDAY	Breastmilk or infant formula feed	Iron-fortified infant wholegrain cereal with pureed banana; formula, breastmilk or cooled boiled tap water can be mixed with the cereal Smooth peanut butter	Pureed fruit, such as banana, kiwi, cooked apple and/or cooked pear; full-fat yoghurt **Introduce a small amount of pureed cooked prawn** Drinks: breastmilk, infant formula, cooled boiled tap water (as required)	Pureed steamed carrot; hummus dip Drinks: breastmilk, infant formula, cooled boiled tap water (as required)	Pureed steamed carrot and mashed potato; steamed boneless fish	Breastmilk or infant formula feed
TUESDAY	Breastmilk or infant formula feed	Iron-fortified infant wholegrain cereal with pureed sweet potato; formula, breastmilk or cooled boiled tap water can be mixed with the cereal Wheat-based porridge or soft Weetabix	Pureed fruit, such as banana, kiwi, cooked apple and/or cooked pear; full-fat yoghurt **Introduce a small amount of pureed cooked prawn** Drinks: breastmilk, infant formula, cooled boiled tap water (as required)	Pureed steamed carrot; mashed egg Drinks: breastmilk, infant formula, cooled boiled tap water (as required)	Pureed beef; pureed cooked carrot, cauliflower and spinach (not mixed together)	Breastmilk or infant formula feed

WEEK 7–INTRODUCING SHELLFISH

	Early feed	Breakfast/ morning snack	Lunchtime	Afternoon snack	Dinner	Bedtime
WEDNESDAY	Breastmilk or infant formula feed	Iron-fortified infant wholegrain cereal with pureed eggplant; formula, breastmilk or cooled boiled tap water can be mixed with the cereal Smooth peanut butter	Pureed fruit, such as banana, kiwi, cooked apple and/or cooked pear; full-fat yoghurt **Introduce a small amount of pureed cooked prawn** Drinks: breastmilk, infant formula, cooled boiled tap water (as required)	Pureed steamed carrot; hummus dip Drinks: breastmilk, infant formula, cooled boiled tap water (as required)	Steamed bone-less fish with pureed steamed carrot	Breastmilk or infant formula feed
THURSDAY	Breastmilk or infant formula feed	Iron-fortified infant wholegrain cereal with pureed pear; formula, breastmilk or cooled boiled tap water can be mixed with the cereal Wheat-based porridge or soft Weetabix	Pureed fruit, such as banana, kiwi, cooked apple and/or cooked pear; full-fat yoghurt **Introduce a small amount of pureed cooked prawn** Drinks: breastmilk, infant formula, cooled boiled tap water (as required)	Pureed steamed carrot; mashed egg Drinks: breastmilk, infant formula, cooled boiled tap water (as required)	Pureed cooked chicken with pureed sweet corn and mashed potato	Breastmilk or infant formula feed

WEEK 7—INTRODUCING SHELLFISH

	Early feed	Breakfast/ morning snack	Lunchtime	Afternoon snack	Dinner	Bedtime
FRIDAY	Breastmilk or infant formula feed	Iron-fortified infant wholegrain cereal with creamed vegetables; formula, breastmilk or cooled boiled tap water can be mixed with the cereal Smooth peanut butter	Pureed fruit, such as banana, kiwi, cooked apple and/or cooked pear; full-fat yoghurt Drinks: breastmilk, infant formula, cooled boiled tap water (as required)	Pureed steamed carrot; hummus dip Drinks: breastmilk, infant formula, cooled boiled tap water (as required)	Steamed bone-less fish and vegetable stew	Breastmilk or infant formula feed
SATURDAY	Breastmilk or infant formula feed	Iron-fortified infant wholegrain cereal with pureed mango; formula, breastmilk or cooled boiled tap water can be mixed with the cereal Wheat-based porridge or soft Weetabix	Pureed fruit, such as banana, kiwi, cooked apple and/or cooked pear; full-fat yoghurt Drinks: breastmilk, infant formula, cooled boiled tap water (as required)	Pureed steamed carrot; mashed egg Drinks: breastmilk, infant formula, cooled boiled tap water (as required)	Root vegetable soup; a small amount of pureed cooked prawn	Breastmilk or infant formula feed

WEEK 7–INTRODUCING SHELLFISH

	Early feed	Breakfast/ morning snack	Lunchtime	Afternoon snack	Dinner	Bedtime
SUNDAY	Breastmilk or infant formula feed	Iron-fortified infant wholegrain cereal with pureed sweet potato and/ or apple; formula, breast-milk or cooled boiled tap water can be mixed with the cereal	Pureed fruit, such as banana, kiwi, cooked apple and/or cooked pear; full-fat yoghurt Drinks: breastmilk, infant formula, cooled boiled tap water (as required)	Pureed steamed carrot Drinks: breastmilk, infant formula, cooled boiled tap water (as required)	Baby ratatouille with pureed cooked lamb	Breastmilk or infant formula feed

WEEK 8–INTRODUCING SOY

	Early feed	Breakfast/ morning snack	Lunchtime	Afternoon snack	Dinner	Bedtime
MONDAY	Breastmilk or infant formula feed	Iron-fortified infant wholegrain cereal with pureed banana; formula, breastmilk or cooled boiled tap water can be mixed with the cereal Smooth peanut butter	Pureed fruit, such as banana, kiwi, cooked apple and/or cooked pear; full-fat yoghurt **Introduce a small amount of pureed cooked soybeans*** Drinks: breastmilk, infant formula, cooled boiled tap water (as required)	Pureed steamed carrot; hummus dip Drinks: breastmilk, infant formula, cooled boiled tap water (as required)	Pureed steamed carrot and mashed potato; steamed boneless fish	Breastmilk or infant formula feed
TUESDAY	Breastmilk or infant formula feed	Iron-fortified infant wholegrain cereal with pureed sweet potato; formula, breastmilk or cooled boiled tap water can be mixed with the cereal Wheat-based porridge or soft Weetabix	Pureed fruit, such as banana, kiwi, cooked apple and/or cooked pear; full-fat yoghurt **Introduce a small amount of pureed cooked soybeans** Drinks: breastmilk, infant formula, cooled boiled tap water (as required)	Pureed steamed carrot; mashed egg Drinks: breastmilk, infant formula, cooled boiled tap water (as required)	Pureed cooked carrot, cauliflower and spinach (not mixed together); full-fat cheese sticks; pureed cooked prawn	Breastmilk or infant formula feed

* Dried soybeans, soaked, boiled and pureed, or else mashed tofu

WEEK 8—INTRODUCING SOY

	Early feed	Breakfast/ morning snack	Lunchtime	Afternoon snack	Dinner	Bedtime
WEDNESDAY	Breastmilk or infant formula feed	Iron-fortified infant wholegrain cereal with pureed banana; formula, breastmilk or cooled boiled tap water can be mixed with the cereal Smooth peanut butter	Pureed fruit, such as banana, kiwi, cooked apple and/or cooked pear; full-fat yoghurt **Introduce a small amount of pureed cooked soybeans** Drinks: breastmilk, infant formula, cooled boiled tap water (as required)	Pureed steamed carrot; hummus dip Drinks: breastmilk, infant formula, cooled boiled tap water (as required)	Steamed boneless fish with pureed steamed carrot	Breastmilk or infant formula feed
THURSDAY	Breastmilk or infant formula feed	Iron-fortified infant wholegrain cereal with pureed pear; formula, breastmilk or cooled boiled tap water can be mixed with the cereal Wheat-based porridge or soft Weetabix	Pureed fruit, such as banana, kiwi, cooked apple and/or cooked pear; full-fat yoghurt **Introduce a small amount of pureed cooked soybeans** Drinks: breastmilk, infant formula, cooled boiled tap water (as required)	Pureed steamed carrot; mashed egg Drinks: breastmilk, infant formula, cooled boiled tap water (as required)	Pureed cooked prawn with pureed sweet corn and potato	Breastmilk or infant formula feed

WEEK 8—INTRODUCING SOY

	Early feed	Breakfast/ morning snack	Lunchtime	Afternoon snack	Dinner	Bedtime
FRIDAY	Breastmilk or infant formula feed	Iron-fortified infant wholegrain cereal with creamed vegetables; formula, breastmilk or cooled boiled tap water can be mixed with the cereal Smooth peanut butter	Pureed fruit, such as banana, kiwi, cooked apple and/or cooked pear; full-fat yoghurt Drinks: breastmilk, infant formula, cooled boiled tap water (as required)	Pureed steamed carrot; hummus dip Drinks: breastmilk, infant formula, cooled boiled tap water (as required)	Root vegetable soup; steamed boneless fish	Breastmilk or infant formula feed
SATURDAY	Breastmilk or infant formula feed	Iron-fortified infant wholegrain cereal with pureed mango; formula, breastmilk or cooled boiled tap water can be mixed with the cereal Wheat-based porridge or soft Weetabix	Pureed fruit, such as banana, kiwi, cooked apple and/or cooked pear; full-fat yoghurt Drinks: breastmilk, infant formula, cooled boiled tap water (as required)	Pureed steamed carrot; mashed egg Drinks: breastmilk, infant formula, cooled boiled tap water (as required)	Pureed roast chicken and vegetable stew; **a small amount of pureed cooked soybeans**	Breastmilk or infant formula feed

WEEK 8—INTRODUCING SOY

	Early feed	Breakfast/ morning snack	Lunchtime	Afternoon snack	Dinner	Bedtime
SUNDAY	Breastmilk or infant formula feed	Iron-fortified infant wholegrain cereal with pureed sweet potato and/ or apple; formula, breast-milk or cooled boiled tap water can be mixed with the cereal	Pureed fruit, such as banana, kiwi, cooked apple and/or cooked pear; full-fat yoghurt Drinks: breastmilk, infant formula, cooled boiled tap water (as required)	Pureed steamed carrot Drinks: breastmilk, infant formula, cooled boiled tap water (as required)	Baby ratatouille with pureed cooked prawn	Breastmilk or infant formula feed

WEEK 9—INTRODUCING TREE NUT

	Early feed	Breakfast/ morning snack	Lunchtime	Afternoon snack	Dinner	Bedtime
MONDAY	Breastmilk or infant formula feed	Iron-fortified infant wholegrain cereal with pureed banana; formula, breastmilk or cooled boiled tap water can be mixed with the cereal Smooth peanut butter	Pureed fruit, such as banana, kiwi, cooked apple and/or cooked pear; full-fat yoghurt **Introduce a small amount of tree-nut butter** Drinks: breastmilk, infant formula, cooled boiled tap water (as required)	Pureed steamed carrot; hummus dip Drinks: breastmilk, infant formula, cooled boiled tap water (as required)	Pureed steamed carrot and mashed potato; steamed bone-less fish	Breastmilk or infant formula feed
TUESDAY	Breastmilk or infant formula feed	Iron-fortified infant wholegrain cereal with pureed sweet potato; formula, breastmilk or cooled boiled tap water can be mixed with the cereal Wheat-based porridge or soft Weetabix	Pureed fruit, such as banana, kiwi, cooked apple and/or cooked pear; full-fat yoghurt **Introduce a small amount of tree-nut butter** Drinks: breastmilk, infant formula, cooled boiled tap water (as required)	Pureed steamed carrot; mashed egg Drinks: breastmilk, infant formula, cooled boiled tap water (as required)	Pureed cooked carrot, cauli-flower and spinach (not mixed together); full-fat cheese sticks; pureed cooked prawn	Breastmilk or infant formula feed

WEEK 9—INTRODUCING TREE NUT

	Early feed	Breakfast/ morning snack	Lunchtime	Afternoon snack	Dinner	Bedtime
WEDNESDAY	Breastmilk or infant formula feed	Iron-fortified infant wholegrain cereal with pureed banana; formula, breastmilk or cooled boiled tap water can be mixed with the cereal Smooth peanut butter	Pureed fruit, such as banana, kiwi, cooked apple and/or cooked pear; full-fat yoghurt **Introduce a small amount of tree-nut butter** Drinks: breastmilk, infant formula, cooled boiled tap water (as required)	Pureed steamed carrot; hummus dip Drinks: breastmilk, infant formula, cooled boiled tap water (as required)	Steamed bone-less fish with pureed steamed carrot	Breastmilk or infant formula feed
THURSDAY	Breastmilk or infant formula feed	Iron-fortified infant wholegrain cereal with pureed pear; formula, breastmilk or cooled boiled tap water can be mixed with the cereal Wheat-based porridge or soft Weetabix	Pureed fruit, such as banana, kiwi, cooked apple and/or cooked pear; full-fat yoghurt **Introduce a small amount of tree-nut butter** Drinks: breastmilk, infant formula, cooled boiled tap water (as required)	Pureed steamed carrot; mashed egg Drinks: breastmilk, infant formula, cooled boiled tap water (as required)	Pureed cooked chicken with pureed sweet corn and potato; pureed cooked soybeans	Breastmilk or infant formula feed

WEEK 9—INTRODUCING TREE NUT

	Early feed	Breakfast/ morning snack	Lunchtime	Afternoon snack	Dinner	Bedtime
FRIDAY	Breastmilk or infant formula feed	Iron-fortified infant wholegrain cereal with creamed vegetables; formula, breastmilk or cooled boiled tap water can be mixed with the cereal Smooth peanut butter	Pureed fruit, such as banana, kiwi, cooked apple and/or cooked pear; full-fat yoghurt Drinks: breastmilk, infant formula, cooled boiled tap water (as required)	Pureed steamed carrot; hummus dip Drinks: breastmilk, infant formula, cooled boiled tap water (as required)	Steamed bone-less fish and vegetable stew	Breastmilk or infant formula feed
SATURDAY	Breastmilk or infant formula feed	Iron-fortified infant wholegrain cereal with pureed mango; formula, breastmilk or cooled boiled tap water can be mixed with the cereal Introduce a small amount of tree-nut butter	Pureed fruit, such as banana, kiwi, cooked apple and/or cooked pear; full-fat yoghurt; pureed cooked prawn Drinks: breastmilk, infant formula, cooled boiled tap water (as required)	Pureed steamed carrot; mashed egg Drinks: breastmilk, infant formula, cooled boiled tap water (as required)	Root vegetable soup; pureed cooked soybeans	Breastmilk or infant formula feed

WEEK 9—INTRODUCING TREE NUT

	Early feed	Breakfast/ morning snack	Lunchtime	Afternoon snack	Dinner	Bedtime
SUNDAY	Breastmilk or infant formula feed	Iron-fortified infant wholegrain cereal with pureed sweet potato and/ or apple; formula, breast-milk or cooled boiled tap water can be mixed with the cereal Wheat-based porridge or soft Weetabix	Pureed fruit, such as banana, kiwi, cooked apple and/or cooked pear; full-fat yoghurt; pureed cooked soybeans Drinks: breastmilk, infant formula, cooled boiled tap water (as required)	Pureed steamed carrot Drinks: breastmilk, infant formula, cooled boiled tap water (as required)	Baby ratatouille with pureed cooked prawn	Breastmilk or infant formula feed

CHAPTER 10
ALL ABOUT "BIOTICS"

THE IMPACT OF ANTIBIOTICS ON THE GUT

It's well known that antibiotic use during the first year of a baby's life increases their risk of allergic diseases. A good study to illustrate this was carried out by the National Center for Child Health and Development in Tokyo, Japan, a city where there has been widespread use of antibiotics. The study found that exposure to antibiotics in the first two years of life increased the risk of children developing asthma, eczema and hay fever at five years of age.

This makes sense, because antibiotics affect the microbiota of young children. Researchers from the University of Helsinki studied the stool samples of 142 children between the age of two and seven attending day care at the time of the study. They found that the use of a particular type of antibiotic called a macrolide (often taken for respiratory infections) led to less stable and less diverse gut bacterial communities. This is not really surprising given what we know about antibiotics. What is a surprise is that these changes appear to last quite a long time. The helpful bacteria from certain groups, as well as the overall richness and maturity of the gut microbiota, remained reduced for up to two years after receiving the macrolide antibiotic.

Of course, it's not always possible to avoid antibiotics in early life, and it's extremely important to remember that good gut health is one thing and safe medical practices are another. Often there is no choice—for example, antibiotic use during pregnancy is standard in most Western countries for women planning to have a C-section. Before the procedure, the mother is given an intrapartum antibiotic prophylaxis (IAP) to help reduce the risk of C-section–related infections, such as endometritis, urinary-tract infection and surgical-site infection.

Certainly, if there is a good medical reason for your baby to be exposed to antibiotics early in life, then this should absolutely be supported. But given what we know about the effects of antibiotics on gut microbiota and the risk of childhood allergy, we should be cautious about inappropriate antibiotic use. Does your baby really need an antibiotic for the common cold? A discussion with your family doctor about this issue is a great place to start.

PROBIOTICS

Probiotics are live bacteria; when you take the right amount of them, they provide a health benefit. Many yoghurts and other fermented foods claim to contain probiotics, but usually this isn't the case—this label is reserved for products that have proven in clinical studies that they have an adequate number of microbes which can confer health benefits.

Probiotics come in many different forms. The most common type is a tablet or capsule. Probiotics can be added to yoghurt, fermented milk and even cheese. Freeze-dried forms of probiotics can be packaged without the need for refrigeration. Probiotics can be made as a gel for vaginal use or as a cream to be applied to the skin.

I'd recommend probiotics be used by pregnant women at high risk of having children with allergies, breastfeeding mothers with a baby at high risk of allergies, and by infants with a high risk of developing allergies.

As a gastroenterologist, I am asked about probiotics all the time by patients. One of the most common questions is: "How long should I take the probiotic?" The answer to this really depends on the clinical reason for why you're taking the probiotic. If probiotics are being used to treat antibiotic-associated diarrhea, then you would only take them for a short period (for the duration of symptoms and a couple of days afterwards). For a chronic condition, such as ulcerative colitis or irritable bowel syndrome, probiotics can be used for a longer timeframe. The effects of probiotics on the gut appear to be temporary, and they disappear when you stop taking the probiotic. We can detect the presence of probiotic bacteria in stools; once a probiotic is stopped, the bacterial presence reduces fairly quickly.

The other question I get asked a lot is: "Which probiotic should I use?" The sheer range of probiotics out there makes it a daunting task to choose the right one, so it's a good question to ask but a difficult one to answer. There are differences between the strains of bacteria used in clinical studies, and drastic differences between the duration and dosage of probiotic use in these studies. Quite often, probiotic strains are mixed together or even combined with other compounds, such as prebiotics. (We'll talk more about prebiotics in the next section.)

The information from well-designed clinical trials is very useful, but there is a saying that 99 percent of statistics only tell 49 percent of the story. As a clinician, I am always cautious about blindly applying the results of studies, even from a meta-analysis. This is because my patient may have individual circumstances that are quite different from those of the population being studied in a clinical trial. For example, a child's ethnicity or their diet may not match the people who were studied in a clinical trial.

Having said that, my advice to patients is always influenced by their specific clinical situation. I will find the appropriate research data on probiotic use for their particular condition, and suggest that, based upon the relevant clinical studies, a particular probiotic strain *may be* of benefit. I will always caution my patient that it may not be helpful for their particular condition, and in no way should it replace standard medical treatment. Nor should we be taking probiotics indiscriminately—a 2019 review of 45 probiotic studies found that there was no benefit to probiotics for healthy people.

I see the role of probiotics as complementary to standard medical care. Like vitamins, probiotics are used widely, and are now part of a multibillion-dollar industry. Many "probiotics" are actually functional foods; they do not come under the same regulations to which drugs, for example, would be subjected. There has been a lot of research carried out to evaluate the effectiveness of probiotics. When it comes to making claims that probiotics can improve the health of children—and especially infants—the benchmark is much higher and more stringent. Companies need to prove in rigorous clinical trials that their product actually works for kids.

What specific allergies might probiotics prevent?

In 2015, a meta-analysis was designed to help inform the World Allergy Organization (WAO) on recommendations regarding probiotic use for preventing childhood allergies. The meta-analysis found that probiotics reduced the risk of eczema when used by women during the last trimester of pregnancy, breastfeeding mothers and infants. There was no benefit observed for other allergic conditions, such as asthma or hay fever. However, other researchers felt that different outcomes needed to be evaluated when assessing how good a probiotic is in reducing allergy risk, not just preventing eczema.

Another meta-analysis published a year later, in 2016, evaluated the ability of probiotics to effectively prevent atopy (the genetic tendency to develop allergies) and food allergies. The meta-analysis found that there was a significant reduction in atopy and food allergies if probiotics were given to *both* the woman during pregnancy and to the infant after they'd been born. When probiotics were only given to the pregnant woman or to the infant, there was no benefit observed.

It may be that probiotics are especially beneficial for infants born via C-section. One Finnish study followed over 1000 infants.

A mixture of four probiotic bacterial strains was taken by the mothers from 35 weeks' gestation, and the infants from birth to six months of age. The children were followed until the age of five, with researchers checking for the development of allergic diseases. Only children delivered by C-section showed any allergy-preventative benefit.

Infants from this study had their stools collected at three months old to assess the composition of their gut microbiota. The research team was particularly interested to see if probiotic use by mothers (starting from 35 weeks' gestation) and their infants could improve the gut microbiota of infants born by C-section or given antibiotics. Their hunch was correct—the probiotic mixture was able to restore normal microbiota composition and function at three months in infants treated with antibiotics or delivered via C-section.

Can probiotics lead to a long-term reduction in childhood allergy risk?

Many of the studies we've looked at have only followed young children for a few years. But thanks to two influential studies, we do have evidence of a long-term benefit in reducing allergy risk from the use of a probiotic in early life.

The first was a Finnish study that followed the infants born to a group of 159 mothers who were randomly assigned either two capsules of a probiotic or two capsules of a placebo. They started taking these capsules four weeks before their due date, and continued to take them for six months after their baby was born, if they were breastfeeding. Babies who were not breastfed were given the capsule contents mixed with water until six months of age.

The children were followed until they were seven and monitored for the development of allergies. The researchers found that the overall risk of developing eczema during the first seven years of life was significantly decreased in the probiotic group, but there was no benefit found for asthma or hay fever. Their conclusion was that probiotic supplements are helpful for preventing eczema in children, but not other allergies.

A 2018 New Zealand study examined the long-term effects of early probiotic use on childhood allergy risk. Pregnant women were randomly assigned at 35 weeks' gestation to take one of two probiotic strains (HN001 or HN019) or a placebo. Women continued taking the capsules until six months after giving birth or the end of breastfeeding, if this occurred earlier. The infants were given capsules daily from birth until two years of age.

Now, remember that the two-year mark is the end of the first 1000 days of life, starting from the time of conception. Clearly, the researchers were thinking about this fact in their study design. This makes the results of the study extremely interesting.

The researchers followed the children to eleven years of age. During their eleventh year, they were assessed for the presence of allergies in the preceding twelve months. What the investigators found was that early-childhood supplements of the probiotic HN001 resulted in children having a 54 percent relative risk reduction in eczema and a 27 percent reduction in hay fever over that twelve-month period. There was no benefit obtained from the other probiotic, HN019.

Over the children's eleven years of life, HN001 was shown to significantly reduce atopic sensitization, eczema and wheeze. HN019 had no effect on these outcomes.

This is a pretty striking finding, because not only does it tell us that a probiotic can make a meaningful difference to allergy risk in the longer term, but it also reinforces the importance of actively and continuously making everyday changes during the first 1000 days. Taking a probiotic strain for much of that first 1000 days of life appears to have a real long-term impact on the risk of eczema *and* other allergies.

This gives hope to those mothers who have infants at a higher risk of allergy because of a family history of allergy, a C-section delivery or exposure to antibiotics during pregnancy/infancy. You can definitely do something to reduce your child's risk of developing allergy, but it's important to be proactive about doing so during those critical first 1000 days of life.

In 2015, the WAO issued guidelines on the prevention of allergies, recommending that probiotics be used by pregnant women with a high risk of allergy in their children, women who breastfeed infants with a high risk of developing allergy, and infants with a high risk of developing allergy. The guidelines note that the main benefit of probiotics is that they reduce the risk of eczema, but as these more recent, long-term studies show, they could have an even greater effect across the allergy spectrum.

WHAT YOU CAN DO AS A PARENT

- Probiotics should be used by pregnant women at high risk of allergies in their children, with the most benefit gained if they start taking probiotics at 35 weeks' gestation.
- Breastfeeding mothers with a baby at high risk of developing allergies should consider taking a daily probiotic until they finish breastfeeding or their baby is six months old, whichever happens first.
- Infants at high risk of developing an allergy should also be given probiotics, and this can be continued safely until two years of age (the end of the first 1000 days of life).
- Advice regarding probiotic choice should be based on the appropriate research data for the particular condition. Dietitians and medical professionals can help provide relevant advice.

PREBIOTICS

If probiotics are the good live bacteria that you're trying to grow in your gut, prebiotics are compounds in food that act as a kind of "fertilizer" for the good gut bacteria. Found in high concentrations in breastmilk, human milk oligosaccharides (HMOs) are complex molecules that act as prebiotics. Unlike the other ingredients in breastmilk, HMOs are not a form of nutrition for the baby but rather for the baby's gut microbes. HMOs have an important role in developing the metabolic and immune systems. They can also stop harmful microbes from sticking to the intestines, which helps to protect a baby from infections.

Galacto-oligosaccharides (GOS) and fructo-oligosaccharides (FOS) are prebiotics that improve the microbiota in your intestines. Together, they are referred to as GOS/FOS. In 2002, scientists first used GOS/FOS to simulate the properties of natural breastmilk. They later found that if they used a specific concentration of GOS/FOS in a mixture, the amount of generated bifidobacteria (an important bacteria for a baby) would be similar to levels typically seen for breastfed infants.

Prebiotics and infant formula

In 2011, the Committee on Nutrition for the European Society for Paediatric Gastroenterology, Hepatology and Nutrition (ESPGHAN) carried out a systematic review to assess the safety and health effects of infant formula supplemented with prebiotics, compared to formula without prebiotics. They determined that giving prebiotic-supplemented formula to healthy infants was safe and caused no negative health effects. But they didn't find evidence of a significant benefit and therefore did not support giving prebiotic-supplemented formula to babies.

The systematic review only covered prebiotics that were already in infant formula and not prebiotics that were added later or given to babies as solids. It's important to note that dietary fiber is actually a prebiotic, and it has important health benefits. Giving your child dietary fiber as part of a balanced diet will absolutely be good for them.

In 2016, the World Allergy Organization (WAO) reviewed all the current evidence and recommended that prebiotic supplements be given to non-exclusively breastfed infants, whether they were at high or low risk of developing allergy. They suggested that doctors and parents *should not* give prebiotic supplements to exclusively breastfed infants, as they felt that these infants would not benefit from prebiotics. The WAO did not make a recommendation about whether or not pregnant women and breastfeeding mothers should use prebiotic supplements. They also didn't make a recommendation about how prebiotics should be given to non-exclusively breastfed babies, but it would make sense for them to be given through infant formula if they're less than six months old. At around six months of age, the baby should be receiving prebiotics mainly through solid foods. The WAO acknowledged that there is a low certainty of evidence in their recommendations and that more research is needed.

At the time of writing, there is a large trial underway in France called PREGRALL. It is evaluating the effectiveness of giving GOS/FOS to 376 mothers during pregnancy in order to reduce the risk of eczema in their babies. Participating women must have asthma, hay fever, eczema and/or food allergy, which puts their infants in the high-risk allergy category. The women in the study will take either the prebiotic or a placebo daily from 20 weeks' gestation to delivery, and their babies will be assessed for eczema at one year of age. It's the first clinical trial to assess the effects of prebiotics taken exclusively during pregnancy in an attempt to prevent eczema in high-risk children. The results will be eagerly awaited and, if positive, there will very likely be updated recommendations in medical guidelines.

WHAT YOU CAN DO AS A PARENT

- It's worthwhile to give prebiotic supplements to non-exclusively breastfed infants.
- If your baby is on infant formula, check whether the formula contains prebiotics.
- If your baby is around six months old, they should be receiving their prebiotics mainly through dietary fiber.
- Prebiotic supplements offer no benefit to exclusively breastfed infants.

SYNBIOTICS

Synbiotics are a combination of both probiotics and prebiotics in a dietary supplement, and they can be either complementary or synergistic. In a complementary synbiotic, the prebiotic and probiotic each offer an independent health benefit. Adding the helpful bifidobacterium to the resistant starch in yoghurt is a good example of a complementary synbiotic. This has been shown in a number of trials to be helpful in treating eczema for children one year and above, but there is no strong evidence that it *prevents* childhood eczema. On the other hand, a synergistic synbiotic is where both the prebiotic and probiotic help each other; usually, the prebiotic acts as a "fertilizer" for the probiotic bacteria.

In addition to eczema, there has been a lot of interest from infant-formula companies in the prevention of food allergies—specifically cow's milk allergy—based on the concept of the atopic march (whereby there is a recognizable and predictable progression in the development of allergies). Both eczema and food allergies are early signifiers of allergic disease, with asthma and hay fever occurring later on. It makes sense to focus on eczema and food allergies first and then extend clinical trials to the other conditions later.

Synbiotics have become increasingly integrated into modern infant formulas, particularly after the recommendation from WAO that prebiotic supplements should be added to the diet of all non-exclusively breastfed infants. But since formula forms a large part of the diet of non-exclusively breastfed infants in the first few months of life it is very difficult to test the effect of synbiotics, because any randomized controlled trial would involve getting rid of synbiotics from the formula.

However, there is an ideal testing environment to assess the effectiveness of synbiotics, brought about by cow's milk allergies. Babies with this allergy cannot drink cow's milk or standard infant formula, but they can have amino-acid formula. Studies are being conducted with amino-acid formula to see if it can be combined with synbiotics to help with the treatment of cow's milk allergy.

The multicenter ASSIGN study assessed 71 infants with suspected non-IgE mediated cow's milk allergy, randomly assigning the babies an amino-acid formula with synbiotics or an amino-acid formula without synbiotics. Impressively, stool samples collected from two-month-old babies who had been receiving the amino-acid formula with synbiotics had a fecal microbiota pattern similar to that of age-matched breastfed infants.

The randomized double-blind PRESTO study was completed in 2021. This study assessed whether synbiotics, when added to amino-acid formula for infants with a cow's milk allergy, would help the babies to tolerate cow's milk earlier. The study found that synbiotics had no effect on babies' tolerance to cow's milk, although they were considered safe to use.

POSTBIOTICS

If prebiotics are "fertilizer" for your gut microbes, then postbiotics are the "waste" caused by that process. Postbiotics can provide plenty of health benefits: they boost the immune system, reduce hypertension and cholesterol, and exert anti-inflammatory and antioxidant effects. A good source of postbiotics is fermented milk, such as yoghurt or kefir.

A fascinating study by investigators in Italy looked at the size of the thymus in newborns. The thymus is a small but special gland located in the front part of your chest, directly behind your breastbone (sternum) and between your lungs. It is instrumental in the production of special immune cells that are critical for the body's defense against viruses and infections.

The size of a baby's thymus depends on how the baby is fed. A breastfed baby's thymus is an incredible twice the size of a formula-fed baby's thymus at four months. This is probably due to the large number of immune components in breastmilk that stimulate the baby's immune development.

In a 2007 Italian study, 30 newborns were exclusively breastfed while the remaining 60 were randomly assigned to either fermented formula or standard formula. Although the thymus of the breastfed babies remained the largest, a fermented formula-fed infant had a significantly larger thymus than a standard formula-fed infant. This raised the possibility that there may be something in the fermented milk that stimulates a baby's immune system.

Fermented formula has also been shown to reduce the severity of acute diarrhea in infants, and to reduce colic. While it has not been shown to reduce the incidence of cow's milk allergy in babies, fermented formula does seem to decrease the degree of sensitivity to the allergen (in other words, a larger amount of cow's milk allergen can be tolerated by allergic babies before a reaction occurs). Postbiotics have huge potential to prevent allergies, but more research is needed.

CHAPTER 11
ENVIRONMENTAL FACTORS

IMPACT OF PETS ON GUT HEALTH

For a long time, many people believed that pets were a risk factor for developing allergies, but a number of recent studies have shown that this is not the case; on the contrary, contact with pets in early life may actually prevent allergies. In 1999, researchers from the University of Gothenburg in Sweden published the results of their research that looked at the risk factors for allergy in children living along the Swedish west coast. Using interviews and skin-prick tests, they were able to show that children in families that had a cat or a dog during the child's first year of life had less hay fever at age seven to nine years and less asthma at age twelve to thirteen years, compared to children with no pets.

In a 2015 Danish study, stool samples were collected from 114 infants at nine and eighteen months old. Microbial analysis revealed that the presence of older siblings and furry pets such as cats, dogs and rabbits led to an increase in bacterial diversity at eighteen months. Other studies have had differing results when looking at the impact of pet ownership on the baby gut microbiota, so more research is needed. One study did find that pet exposure was very effective in reducing the tendency for allergy in children delivered by C-section, suggesting that these kids are the ones who would benefit the most from keeping a household pet.

At a societal level, it appears that in communities where there is a high prevalence of cats and cat allergens, having a cat in the home does not bring about an increased risk of allergic disease. And where the prevalence of cats and cat allergens is low, having a cat at home can increase the risk of allergy. It seems quite contradictory that exposure to a high amount of cat dander, which is thought to trigger allergies, can actually help prevent allergies. But what's happening is a process called immune tolerance.

In 1829, the first written record of oral immune tolerance appeared in a medical journal report describing how an indigenous South American tribe would chew poison-ivy leaves to prevent bad skin reactions when they came into contact with the plant. In 1911, scientist and author H.G. Wells carried out the first experimental research in oral immune tolerance, when he and a colleague, biochemist Dr. Thomas Burr Osborne, showed that guinea pigs which had already developed anaphylaxis after eating egg white did not experience more anaphylactic episodes from egg white if they were fed small amounts of egg white over time.

Nineteen years after their first study, in 2018 the same researchers from the University of Gothenburg came up with the concept of the "mini-farm" immune environment. This is the idea that exposure to pets in early life can lead to tolerance not only to the pet itself but also to food and airborne allergens more broadly. Intriguingly, they found that the prevalence of allergic diseases in children aged seven to nine years was reduced exponentially based on the number of cats and dogs in the home during the first year of life. The more cats and dogs a household kept during that first year, the greater the protective effect on the children later on. They used the term "mini-farm" to refer to this allergy-protective effect.

Of course, there are other protective factors at work in early childhood. The University of Gothenburg researchers felt that exposure of a young child to a dog or a cat may have a protective effect in children. However, if the child already has other protective factors, such as older siblings, then a dog or cat may not provide any extra protection—unless the child is exposed to several animals, hence the "mini-farm" concept.

The researchers speculated that immune tolerance through a "mini-farm" effect in these children comes about through close contact with the animals. This means that, rather than just the allergen (dander of the cat or dog) making contact with the child, microbes from the animals, along with their endotoxins (which are released when bacteria die), had to be getting to that child. The microbes and endotoxins were probably entering the child through the fecal-oral route; that is, via some of the animal's poo getting into your child's mouth and then into their body. Even though this sounds disgusting, it turns out that poo has some features that can help to reduce allergy risk. See Chapter 12 for more information on this.

A number of studies have proven that growing up on a farm can provide some form of protection against allergies. Farm environments such as stables or other areas where livestock and poultry are kept contain especially high levels of endotoxins. So it stands to reason that environmental exposure to the endotoxins and microbes found in rural locations could help protect children from developing allergies.

As we discussed in Chapter 3, researchers believe that early exposure to endotoxins can skew a Th2 response to a Th1 response. However, it has also been discovered that, in patients with *established* asthma and hay fever, exposure to endotoxins later in life can actually aggravate their allergic symptoms. A study carried out on a rat asthma model by researchers from The University of Western Australia demonstrates just how important the timing of the exposure is: if a rat was given a dose of endotoxin within the first four days of being sensitized to an egg white, they did not go on to develop asthma. But if they received the endotoxin six or more days after the egg-white allergen, the rats developed asthma symptoms. So in terms of building immune tolerance, it seems that the earlier the exposure to endotoxins, the better.

We used to think that owning household pets led to children being exposed to a higher level of allergens, and that living in the city resulted in a greater exposure to allergens. This could not be further from the truth. Research tells us that allergens from pollens, molds and animals are actually much greater in rural than urban areas. There is also much greater and more frequent contact between kids and pet allergens in country homes, but this doesn't seem to lead to an increased risk of allergy. In fact, as we've seen in studies of the children of farmers, the opposite seems to be true.

THE BENEFIT OF FARMS

There is something quite special about the farm environment. Kids who regularly drink farm milk have an increased number of immune cells that regulate their immune responses. Kids living on farms have a different gene expression of what are called "toll-like receptors," which are part of the first line of defense of the immune system. They are sentinel watchtowers that can send out a strong message when they encounter invading microbes. The strength of the signal that they emit is important. Weak toll-like receptor signaling is linked to allergic asthma, while strong signaling protects against asthma. One systematic review examined the data from 39 studies and calculated that children living on a farm had approximately 25 percent lower asthma prevalence.

The thinking is that city kids simply haven't grown up with the mature and diverse microbial and allergen exposure that is protective. But the good news is that it's not too late—even for adults—to be exposed to the farm environment and receive some protective benefits. While some studies have shown that children growing up in an urban environment might have their pre-existing allergies worsened by being exposed to the farm environment, a 2013 Danish study showed that exposure to a farm environment protects against the development of allergies not only in childhood but also in young adulthood. Regardless of whether or not you were exposed to rural environments as a child, working on a farm as an adult has a protective effect against common allergies.

However, I wouldn't leave it too long if you're planning to take your children to a farm. An interesting study was published in 2017, comparing the immune development of the Amish and Hutterite children. The Amish and Hutterite populations migrated to the United States from Europe in the 1700s and 1800s, respectively, because of religious persecution. The Amish settled on single-family farms in Pennsylvania, Ohio and Indiana. The Hutterites settled on three communal farms in South Dakota. While both populations retain traditional lifestyles and a strict interpretation of the Bible, they differ in one very significant way. The Amish eschew all modern technology and practice traditional farming. They keep a variety of animals on their farms, such as cows, horses, poultry and rabbits, which their children are exposed to from an early age. On the other hand, the Hutterites embrace modern technology and have industrial-sized farms. Due to the distance of the Hutterite farms from their homes, Hutterite children younger than six years old are not exposed to farm animals.

Amish farm children are known to have a low prevalence of asthma and allergic sensitization, while Hutterite farm children have a high prevalence of asthma and allergic sensitization. Distinct differences were found between the innate immune responses of the two communities. Researchers reason that the high prevalence of allergy in Hutterite children is due to a lack of access when they are young to the protective exposures of the farm environment. Young city children similarly often lack access to these same allergy-protective exposures typically found in farm environments.

The director of the Center for Clinical Research and Prevention in Denmark, Professor Allan Linneberg, believes that the reason for the rise in allergy in the Western, urbanized world is the reduced exposure to the most common environmental allergens, such as pollens, house dust mites and molds. He emphasizes that this is currently only a hypothesis that warrants further testing, but the data to support this looks promising; if it can be convincingly proven, then it would change the way that we view allergens and the whole concept of allergen avoidance.

We know that exposure to the farm environment reduces allergy risk in children by acting upon the immune system, and now we have more insights into how that process might come about through changes to the infant gut microbiota. In 2019, researchers from the University of Wisconsin collected stool specimens from two-month-old babies born to farm families and non-farm families. The gut microbiota of these infants was found to be quite different. In particular, there were more members of the bacterial class Clostridia in the farm-infant microbiota.

This is an interesting result, because in an American mice study, Clostridia was shown to cause immune cells to produce high levels of interleukin-22, a signaling molecule that controls the passing of material from inside the gut and intestines, through the lining of the gut wall and into the rest of the body. The results of the study concluded that the higher production of interleukin by Clostridia appeared to prevent peanut allergens from getting into the bloodstream of mice. So this study revealed solid evidence about the links between the gut microbiota and the immune system, showing us how environmental influences, such as a farm environment, can change the gut-microbiota composition of babies and alter their immune system and the risk of allergy.

UNPASTEURIZED MILK

We know that there are several elements of farm living that result in young kids being exposed to a wide diversity of microbes. One such factor is the consumption of farm or unpasteurized milk. I was once a guest on a live panel organized by Time Out Sydney, where we discussed the use of unpasteurized milk as part of the topic "Is the nanny state taking the taste from our restaurants?" There were four of us on the panel: a restaurateur/ owner of a few fine-dining restaurants in Sydney, an owner of a large burger joint, an investigative journalist and me. Guess who got to defend the nanny state on that panel, in front of a whole audience of hipster students and hospitality workers? You can imagine the reaction I got when I raised concerns about the safety of consuming farm milk.

I stressed the importance of pasteurization for society. Pasteurization is the technique of treating milk to stop bacterial contamination. By heating milk to 72 degrees Celsius (162°F) for at least fifteen seconds, many harmful bacteria, such as *Salmonella*, *Campylobacter* and *Listeria*, are killed off. Selling raw farm milk to the public is illegal in Australia, and there is a good reason why—young children have died after drinking raw milk. There is evidence that farm milk reduces allergies as it contains a range of bacteria, but I would strongly discourage parents from giving their children raw farm milk.

DAY CARE

We enrolled Brandon in our local day care center for three days a week, when he was eight months old. As we all know, there is a lot of microbial exposure in day care centers and ample opportunities to increase the diversity of the gut microbiota species. A study from North Carolina found exactly that when they compared the gut microbiota of infants attending day care to that of infants cared for at home. Some studies have found that exposure to other children at day care can protect against the development of asthma and wheezing, while other studies have given conflicting results. More research still needs to be done in this area.

WHAT YOU CAN DO AS A PARENT

- During your baby's critical first year of life, consider having a pet in the house or exposing your infants to the household pets of other people. However, refrain from doing this if a family member has a pet allergy.

- Consider taking your family for a getaway on a traditional farm during your baby's first year of life. There is also some evidence to suggest that time spent on farms provides a protective effect even for older children and young adults.

CHAPTER 12

FUTURE TREATMENT STRATEGIES

We're all hoping that one day a vaccine or cure will become available for allergies. There are some interesting treatment strategies being developed—some are experimental; some show a lot of promise. Here are a few that are making promising progress.

VAGINAL SEEDING

What happens if a woman requires a C-section, but she has concerns about the baby developing allergies through lack of exposure to vaginal microbes? A new and controversial approach called vaginal seeding is currently being tested in clinical trials, where newborns delivered by C-section are swabbed with their mother's vaginal fluids soon after birth. How is this performed, and is it safe?

As we've already seen, the gut microbes of C-section infants are very different from those of vaginally delivered infants. Vaginal seeding involves placing a gauze swab in the mother's vagina for one full hour just prior to the C-section delivery. Right before the C-section, the gauze is removed from the vagina and placed in a sterile dish at room temperature. This gauze swab—which contains the mother's vaginal fluids and microbes—is then wiped over the newborn's mouth, face and body within one minute of delivery.

In 2016, a team of researchers from the University of California San Diego examined eighteen infants and their mothers. Seven infants were delivered vaginally, and eleven by C-section. Four out of the eleven C section infants were treated with vaginal seeding immediately after birth. Thirty days later, the four C-section infants who had had vaginal seeding showed a skin and gut microbiota that was more similar to vaginally delivered babies than C-section infants who didn't receive the treatment. The study wasn't designed to show health outcomes, but the researchers believe that their study does show that it's feasible to transfer vaginal microbes to the infant.

The study received widespread attention both from traditional media and the academic community and was critically examined for weaknesses. One of the points made was that all mothers who delivered by C-section received antibiotics, while only one mother who delivered vaginally was exposed to antibiotics. Therefore, it has been argued that changes in the gut microbiota cannot be purely due to the vaginal seeding, as the effects of antibiotics have to be taken into account. There was also a very small sample size, with only four C-section infants being treated. Another valid criticism was the lack of accounting for potential differences in the weight gain of mothers during pregnancy, as this might influence the microbiota results.

Some researchers are concerned that the practice of vaginal seeding is becoming more mainstream, and is being performed without the supervision of health-care professionals. This is a very important point, because there is a real risk of transferring harmful microbes from the mother's vagina to the newborns. These microbes may not be causing any obvious issues for the mother.

A good example of this is Group B streptococcus (Group B strep for short), which is commonly found in the intestines and vagina. It is not considered to be harmful for healthy women and about 20 percent of women will have Group B strep in their vagina around the time of giving birth.

The main concern with Group B strep is passing it on to the baby before or during a vaginal birth. A small proportion of babies (one in 200) who are infected with Group B strep can become seriously ill during the first few days of life with life-threatening infections such as pneumonia or meningitis.

An obvious challenge with vaginal seeding is that even if Group B strep testing is negative at 35 weeks, there is certainly a chance that the mother could be infected by the time of C-section delivery at 40 weeks. Furthermore, Group B strep isn't the only microbe in vaginal fluid that may be risky for the infant. Vaginally transmitted *E. coli* infections can be problematic, and there are always concerns about undiagnosed chlamydia, gonorrhea and herpes simplex virus.

The American College of Obstetricians and Gynecologists (ACOG) reviewed the practice of vaginal seeding. They concluded that it is not recommended and should only be performed in the context of a research study, which I agree with. Vaginal seeding has potential, but at the moment it is much too risky to try it outside the controlled environment of a research program.

IMMUNOTHERAPY

Researchers have long been trying to create an allergy vaccination. Immunotherapy is a type of allergy vaccination whereby the body is exposed to small amounts of an allergen in gradually increasing doses so that eventually the body can build up immunity to the allergen. This means that there should be a reduced response to the allergen when the body is re-exposed in future.

Immunotherapy has been shown to be effective for improving asthma symptoms, and it has a clearly established role in insect sting allergies. Both of these involve injections into the skin. Researchers previously believed that injection-based immunotherapy may also be useful for food allergies and the first attempts at immunotherapy for peanut allergy were made in the 1990s, using peanut extracts injected under the skin. Unfortunately, these had to be stopped because of some severe adverse reactions. Swallowing the allergens was then embraced as the best method of conducting immunotherapy, and this has become known as oral immunotherapy.

However, oral immunotherapy is still experimental. Treatment is individualized, and there are no standard protocols in place at this stage. For peanut allergies, there have been multiple studies with varying results. Sometimes oral immunotherapy is combined with medications, such as omalizumab, which can block the cross linking of IgE (allergy) antibody receptors on immune cells and thereby reduce the severity of an allergic reaction.

An important systematic review and meta-analysis of oral immunotherapy for peanut allergy was published in late April 2019. The meta-analysis decided to focus on the most important concern for people with peanut allergy—having an anaphylactic reaction. The researchers felt that this was a much better measure of how a severe allergy affected a person's life than just a positive result on an oral peanut challenge.

The systematic review included 1041 participants across twelve randomized controlled trials of peanut oral immunotherapy. The meta-analysis found that oral immunotherapy actually increased the risk of anaphylaxis, with peanut oral immunotherapy being associated with 151 more episodes of anaphylaxis per 1000 patients.

Oral immunotherapy may well be helpful for some cases of food allergy, but given the need to properly determine safety and effectiveness, it should only be used in the context of a clinical trial. If your child has a food allergy, complete avoidance of that food is still the best course of action. Educating yourself, your family, your child's carers and your child on what foods contain the allergen, and what to do if your child has an allergic reaction, is vitally important.

FECAL MICROBIOTA TRANSPLANTATION (FMT)

You might have read about or seen coverage of the pretty distasteful but quite fascinating practice of fecal microbiota transplantation (FMT). It involves the transfer of a stool from a healthy donor to the gut of a sick person. Chinese physicians practiced it as far back as 1700 years ago by giving patients with food poisoning or severe diarrhea "yellow soup."

In 1958, Professor Ben Eiseman and his colleagues from Denver Veterans Administration Hospital in the United States used FMT to treat four cases of *Clostridium difficile* colitis (a bacterial infection in the colon). It was given to the extremely unwell patients in the form of enemas. All four had quick recoveries after the FMT, and they were successfully discharged from hospital. FMT is now well established as the treatment for severe *Clostridium difficile* colitis. Administration via colonoscopy is one option, but there are a few other methods of delivery. Patients can swallow fecal matter sealed in capsules (or "crapsules," as they are better known in the media), which is the most common way to receive FMT. It's certainly much less invasive than getting it via colonoscopy or nasogastric tube. Patients generally prefer a capsule to an enema, but they have to overcome the mental "yuck factor" of ingesting feces.

How does FMT work? The answer is that no one really knows. We do know that FMT provides a very high dose of bacteria to the gut, and it increases the diversity of bacterial species in the gut. Donor stools vary from day to day, and they largely reflect the diet and general health of the donor. Unsurprisingly, there is variation between the fecal microbiota of healthy donors. In fact, some donors are known as "super donors," because in clinical studies their stools result in significantly more successful FMT outcomes than the stools of other donors. Due to the nature of clinical trials, these super donors are generally unaware that they are producing something truly special, but they are of great interest to companies that produce commercial FMT capsules.

FMT and allergies

One interesting finding about FMT supports its potential use for preventing allergies. In an experiment on mice with colitis, FMT was shown to exert multiple effects on immune defenses. In particular, FMT resulted in the secretion of interleukin-10, which led to a reduction in gut inflammation. Interleukin-10 not only helps to reduce gut inflammation but can also damp down the allergic response.

Researchers from Boston Children's Hospital Division of Allergy and Immunology in the United States transplanted fecal matter from healthy human infants into specially bred mice with egg allergies. The research team found that FMT protected the mice against anaphylaxis. Transplanting fecal matter from babies with food allergies, on the other hand, did not protect the mice from anaphylaxis. The researchers then fed the allergic mice a group of six *Clostridium* species and found that this protected the mice against allergies. These bacteria are found in higher numbers in farm babies and have been shown to prevent peanut allergens from getting into the bloodstream of mice. This research shows, in animals at least, that FMT could help protect against allergies.

But we don't know if FMT can protect against allergies in humans. The same Boston Children's Hospital researchers are investigating the use of FMT for peanut-allergy prevention.

The researchers are planning a pilot study using ten adults aged eighteen to 40 years with severe peanut allergy. The main aim of the study is to assess how safe and tolerable FMT capsules are for this group of patients. All patients must react to a minimum of 100 milligrams of a peanut challenge (less than half a peanut). The investigators will be sourcing the stool samples from the non-profit US stool bank OpenBiome. Donors for the study are to be completely free of peanut and tree-nut allergies. They will also need to pass an extensive screening test.

Eligible patients in the study will receive two days of FMT via capsules. Four weeks later, and then every four months afterwards, they will have a peanut challenge of up to 600 milligrams (2.5 peanuts). The study is due to be completed in mid-2021. If the results prove successful, then the investigators will try to see if FMT can prevent peanut allergies in children.

Currently, there are no approved uses for FMT in infants, apart from severe *Clostridium difficile* colitis; even for that condition, there is not a lot of data. The first FMT carried out for a child in their first 1000 days of life was on a thirteen-month-old male infant treated for severe *Clostridium difficile* colitis at Shanghai Children's Hospital.

Doctors followed up with the child for three years after the FMT, and no adverse effects or abnormal changes to his growth or behavior were observed. Since that first case, the Shanghai Children's Hospital Department of Gastroenterology has gone on to treat a four-month-old infant, who is the youngest child treated with FMT that I could find when I trawled through the medical literature. FMT may well have a role to play in the prevention of serious allergies, particularly in very young children, but we need to be careful and await good-quality clinical-trial data.

Complications with FMT

Fecal microbiota transplantation (FMT) is still regarded by regulatory bodies as experimental. However, regulatory bodies do not strictly control it, and this has allowed FMT to be used without difficulty in many countries. This may change soon. The US Food and Drug Administration (FDA) issued a safety alert in early June 2019, following the death of a patient who received FMT. Two other people with weakened immune systems also received FMT from the same donor. They both developed multidrug-resistant *E. coli* infections, and one of them died too.

The FDA now requires that all stool samples used in FMT be tested for drug-resistant microorganisms. All donors will also need to be screened for potential drug-resistant infections. Anyone at higher risk for colonization with multidrug-resistant organisms will be excluded from FMT use. This is a welcome announcement given the widespread differences between institutions in how FMT is delivered, and the lack of long-term data for FMT.

BACTERIAL CONSORTIUM TRANSPLANTATION (BCT)

A safer option than FMT may be bacterial consortium transplantation (BCT). Rather than using feces, which has many different components, BCT comprises only bacteria. Essentially, particular groups of bacteria are chosen from donor fecal microbes and given to the patient. The bacterial combination can be carefully selected so the person only receives non-harmful species, rather than all of the species as in a regular FMT.

A proof-of-principle study was published in 2013, describing how a group of Canadian scientists were able to create a bacterial combination of 33 different purified gut bacteria from a healthy donor. They used this for two patients with *Clostridium difficile* colitis. Both patients displayed a similar response to those who have had FMT, and their bowel movements were normal within two to three days. They were still symptom-free when checked six months after BCT.

PHAGE THERAPY

Given what we know about the importance of gut microbes in preventing the development of allergies in children, it follows that we might think about trying to modify the microbes in their gut. But selectively modifying gut microbiota to reduce our children's risk of allergy is challenging. Antibiotics kill bacteria indiscriminately, and therefore using an antibiotic will get rid of both beneficial and harmful gut bacteria.

A small virus that infects and kills bacteria is called a bacteriophage (or phage for short). Phages have evolved to kill just a specific set of bacteria, and so they could be excellent tools for targeting a harmful bacterial species without disturbing the rest of the beneficial gut bacteria. There is certainly potential for predatory viruses to be used to modify and reshape the gut microbiota to reduce allergy, but we need to take a cautious approach and ensure that there is good experimental data.

Some promising data for phage therapy comes from a limited number of reported cases, in particular related to Netherton syndrome. This is a rare disorder that affects the skin, hair and immune system. Newborns with Netherton syndrome have an extreme form of eczema—red, scaly skin that can leak fluid.

In 2017, doctors in the country of Georgia reported on the use of phage therapy for a sixteen-year-old boy with Netherton syndrome. The boy had very serious skin infections and multiple allergies to antibiotics, and he was frequently admitted to hospital. In desperation, the doctors turned to the use of phage therapy. After seven days of phage therapy, there was a significant improvement in his skin. He continued to receive the therapy and at his six-month follow-up, his condition had stabilized, and his skin was looking a lot better. No allergic reactions to the phage therapy were observed.

SO, WHAT ARE YOU WAITING FOR?

This book began five years ago, as I stood in a Chinese restaurant petrified I was about to lose my baby girl. As a new parent, it was terrifying not knowing what to do, even although I had more than a decade of medical practice under my belt.

I started writing this book from a place of fear. We have all been conditioned to believe that fear is a bad thing and something we should avoid at all costs—especially when dealing with our kids' health. But if you, like me, are a little uncertain and afraid of what life might have in store for your kids then I think you're actually on the right track. It means that you care. It means you want the best possible future for them.

But not every parent is in a position to be able to delve into the root causes of their child's allergies. As an academic and a clinician, I feel really lucky to be able to do so. We're also very fortunate to have reached a moment in history where there's a lot of information available about childhood allergies and new advice for how we can prevent them developing in the first place. I wrote this book to distil and explain all that information for parents, so that rather than being fearful of allergies and their many complications, you're instead empowered to take action to prevent or better treat them.

And you can start making positive changes today. As you now know, the evidence suggests that making very small lifestyle changes can lead to dramatic changes for the health of our kids. By being proactive, you can set up your child's immune system for life.

When your baby is exposed to "old friends," their immune system begins to function normally and they learn how to correctly identify good and bad microbes. The importance of getting your kids out into the natural environment for this process can't be understated—there are so many health benefits from time spent in the outdoors for kids. And this starts from the very beginning, since mothers who visit stables or farms while they're pregnant are actively reducing the risk of allergies in their unborn children. It is a radical concept to embrace, but we now know that interventions during that crucial first 1000 days from conception make all the difference.

We also know that good nutrition and stress relief during pregnancy have far-reaching impacts on babies' health. Parents have long known that breastmilk is important for their babies, but until relatively recently we didn't fully understand its powerful anti-allergenic properties. And where mothers are unable to exclusively breastfeed for the full first six months of their baby's life, prebiotics have come a long way in filling the gap.

Indeed, the use of "biotics" has made us realize that the baby's gut—and particularly the wealth of microbes living in the gut—plays a dynamic role in allergy prevention. Good nutrition is so important for the health of your baby's gut, and feeding your baby a diverse range of foods in the first year of life will protect them from allergies. Studies around the world have supported the introduction of food allergens such as peanut within the first year of life in order to promote tolerance rather than sensitization, with the critical time for the safe introduction of solid-food allergens being between six and twelve months. The nine-week infant meal plan in Chapter 9 provides a clear pathway for you to put this into action, by introducing common allergens gradually and carefully.

There are also a few basic things that are definitely worth thinking about during the first 1000 days of your child's life, such as taking probiotics, introducing dietary fiber to your child's diet early on, and having a pet (or two) living in your home.

If your child has an existing allergy, then there are usually things that can be done to better manage their condition. However, in some cases, complete lifelong avoidance of the allergen is the right course of action. I know that can be tough to hear for parents with kids suffering from severe allergies, but there is so much research going on right now in allergy studies and on the gut in general. It's an evolving field with a growing body of research into exciting innovations, such as fecal microbiota transplantation and phage therapy. These new developments show promise for both the prevention and treatment of allergies, and I feel hopeful for the future.

You've taken the first step by reading this book and I hope you're inspired and empowered to look at baby gut health and childhood allergies in a different way.

One of the most useful things I have ever been taught is the practice of visualization. When I think about what I want my future to look like, I see Olivia and Brandon as safe, healthy and happy kids. It's a pretty simple dream but I'm doing everything I can to achieve it.

Think about what you want for your kids. And then start taking steps to make it happen.

So, what are you waiting for?

GLOSSARY

Adrenaline—a natural hormone produced by the body in response to stress

Allergen—a substance that can cause an allergic reaction

Allergen immunotherapy (AIT)—the practice of giving gradually increasing doses of an allergen to which a person is allergic causing the immune system to become less sensitive to that allergen

Allergic rhinoconjunctivitis—the medical term for hay fever; symptoms affect the eyes as well as the nose

Allergy—an unnecessary immune response to substances in the environment that do not bother most people

Anaphylaxis—a severe, life-threatening allergic reaction

Antibody—a protein produced by the immune system to fight foreign microorganisms

Antihistamine—a drug or other compound that inhibits the physiological effects of histamine; used especially in the treatment of allergies

Asthma—an allergic reaction and respiratory condition causing difficulty breathing

Atopic march—the recognized progression of genetically inherited allergies

Atopy—the genetic tendency to develop allergies

Bacterial consortium transplantation (BCT)—procedure where particular groups of bacteria are selected from donor fecal microbes and given to a recipient

Commensal gut bacteria—the usual microbes of our gut that don't cause harm and are beneficial for our health

Corticosteroids—steroid hormones that are known to lower inflammation in the body and reduce immune system activity

Development phase—this is the first phase of the infant gut microbiota development. In this phase bacteria rapidly colonize the sterile gut and become more numerous and diverse as the baby draws nourishment from food sources e.g. milk and solid foods

Eczema—an allergy whereby the skin becomes rough, inflamed and itchy

Endotoxin—a toxin present in the cell walls of bacteria that is released when the bacteria dies

EpiPen—a portable auto-injector filled with adrenaline, for use in anaphylaxis in adults

EpiPen Junior—a portable auto-injector filled with adrenaline, for use in anaphylaxis in young children

Fecal microbiota transplantation (FMT)—procedure where fecal material including live microbes is transferred from a donor to a recipient

First 1000 days of life—this is the critical time period in a young child's life starting from day zero (conception) to two years of age (1000 days) which shapes their development and well-being

Gut microbiota—the microbial population that lives in our digestive tract

Hay fever—*see* allergic rhinoconjunctivitis

Histamine—a chemical released by the body that causes allergic reactions, such as itchiness and wheezing

Hygiene hypothesis—the theory that allergies arise when a child does not have enough exposure to microbes in early childhood

IgE (allergy) antibodies—antibodies that travel to particular cells which release chemicals, causing an allergic reaction

IgG antibodies—the most common type of antibody in our bodies and protects against infection

Immune response—a reaction which occurs within the body to defend against invaders such as viruses, bacteria and parasites. It can also be triggered if the body is exposed to allergens, cancers and tissue damage

Immune system—complex network of cells and proteins that defends the body against diseases such as infection and cancer

Immune tolerance—a state of unresponsiveness of the immune system to substances that can trigger an immune response in the body

Immunology—the study of the immune system and how it can protect us from disease

Lymphatic system—a network of tissues and organs that help rid the body of toxins, waste and other unwanted materials

Meconium—a newborn's first poo

Meta-analysis—an examination of many scientific studies on a particular subject in order to see trends in data

Metabolic—the breakdown of food and its transformation into energy

Microbes—organisms which are microscopic and thus so small that they cannot be seen with the naked eye

Microbial colonization—the process by which microbes invade and establish a self-sustaining population

Microbial diversity—the range of different types of microbes and their abundance in a community

Microbiome—the community of microorganisms that lives in or on a person, primarily in their gut, including the microbes' genetic material. This term is often used interchangeably with "microbiota"

Microbiota—the community of microorganisms that live in or on a person, primarily in their gut

Microorganisms—a microscopic organism, such as a bacterium or virus; often used interchangeably with "microbes"

Pediatric immunologist—a medical specialist who studies and treats disease of the immune system in children including allergies

Phage—a type of virus that infects bacteria and destroys them

Postbiotics—the metabolic by-products or "waste products" of probiotic bacteria

Prebiotics—compounds in food that act as a "fertilizer" for good bacteria in the human gut

Premastication—practice of feeding an infant with food that has been chewed by their mother or other caregivers

Probiotics—live bacteria that have beneficial qualities when ingested

Randomized controlled trial—a study that randomly assigns participants to either an experimental group or a control group

Sensitize—where a foreign substance is recognized as "abnormal" by the immune system and results in the production of IgE antibodies specific for that allergen. Encountering the same allergen in the future can but does not always trigger an allergic reaction

Skin-prick test—a small plastic pricker is used to make a shallow break in the skin, and a tiny amount of allergen is dropped into the pinprick

Steiner school—a school that bases its educational philosophy on developing learners' intellectual, artistic and practical skills in an integrated and holistic manner

Sterile—a microbe free environment

Synbiotics—a combination of both probiotics and prebiotics in a dietary supplement

Term newborns—an infant born between 37 weeks and 42 weeks of gestation. Also known as "full term"

Th1—the pro-inflammatory response of the immune system that can cause tissue damage

Th2—the anti-inflammatory response of the immune system that counterbalances Th1 but can lead to allergic reactions

Thymus—a small gland located directly behind the breastbone (sternum) and between the lungs. It is instrumental in the production of special immune cells that are critical for the body's defense against viruses and infections

Toll-like receptors—regarded as part of the first line of defense of the immune system they can be considered as "sentinel watchtowers" that are able to send out a strong message when encountering invading microbes

Tongue-thrust reflex—a reflex where the baby sticks its tongue out/forward when its lips are touched or food is placed in its mouth. This reflex helps to protect babies from choking or aspirating on food

Transitional phase—a second transitional shift in the infant gut microbiota which usually occurs between one and two years of age. After the transitional phase the toddler's gut microbiota becomes more similar to an adult and remains fairly stable

Vaginal seeding—a technique where newborns delivered by C-section are swabbed with their mother's vaginal fluids soon after birth

Weaning—the process by which babies who were fully reliant on milk are introduced to solid foods

ACKNOWLEDGMENTS

The acknowledgments are so great to write because I can finally take my foot off the pedal and just start thanking people. And I do want to thank the many people who have been instrumental in the production of this book. For the opportunity to be able to write this passion project of mine, I am grateful to Kylie Westaway, Claire Kingston and Tom Gilliatt from Allen & Unwin. I also want to acknowledge Geraldine Georgeou from Designer Diets, with appreciation for her review of the nine-week infant meal plan.

Books undergo many revisions before they become a reality, and I've certainly had to tweak the content a few times to get it to the version that you've just read. For this I am indebted to my copyeditor, Samantha Sainsbury, who has done a tremendous job in both structural editing and copyediting.

I must give a huge thank you to my editor at Allen & Unwin—Tessa Feggans. Tessa has believed in this book from day one and has worked tirelessly to make the book a reality. I truly could not have done this without her support and I'm forever grateful for the time she spent helping me craft the best version of this book.

I am also very grateful for the assistance of Paula Ayer from Greystone Books in making this version of the book for North America. Her support has made my writing an enjoyable and smooth process.

I want to thank my dear wife, Cindy, as she's the one who had to put up with me through the many late nights and weekends I spent researching and writing this book. That she did so with eternal patience and encouragement is testament to the wonderful and supportive person that she is. Cindy's words and experiences are scattered throughout this book to provide the perspective of a caring mother—she is an amazing mom, and I hope she knows how much I love and appreciate her.

And finally, to my dearest children, Olivia and Brandon—you both inspired me to write this book. You give literal meaning to the first 1000 days of life and I love you both so very much.

INDEX